Stan Rodski is a highly respected scientist and an authority on how to improve mental performance in high-stress situations by adopting brain energy management techniques. A cognitive neuroscientist, he practised as a registered psychologist for over thirty years but now focuses on research, lecturing and writing. His current role as Chief Neuroscientist for The Think Group has led to him working across Australia and internationally to develop innovative technologies and programs for individuals (brain coaching), peak-performance sport and eSport teams, and many top-500 companies looking for creative, brain-focused initiatives to facilitate corporate success. Most recently, he has been applying his research in the brain sciences to issues such as reducing sleep deprivation and brain fatigue and congestion, and managing and boosting energy to harness stress.

Stan introduced the brain-science-based *Colourtation* method to stress management and his work using colouring-in techniques led to his bestselling books, *Anti-stress*, *Brain-science* and *Modern Medi-tation*, which were featured in Oprah Winfrey's 2016 Christmas Wish List.

Stan's other specialist area of research is the neuroscience of machine learning – or AI (artificial intelligence) – particularly using neurologically based predictive algorithms (computational neuroscience) in areas such as diversity and recruitment. Most recently, he has developed innovative 'brain-edge' scans and 'neuro-POD' technologies, which seek to revolutionise human performance in personal, learning and workplace environments.

Stan is married with four children and three grandchildren and lives in Melbourne, Australia.

THE
NEUROSCIENCE
OF MINDFULNESS

THE NEUROSCIENCE OF MINDFULNESS

THE ASTONISHING SCIENCE BEHIND HOW EVERYDAY HOBBIES HELP YOU RELAX, WORK MORE EFFICIENTLY AND LEAD A HEALTHIER LIFE

STAN RODSKI

HarperCollins*Publishers*

First published in Australia in 2019
by HarperCollins*Publishers* Australia Pty Limited
ABN 36 009 913 517
harpercollins.com.au

Copyright © Mind Peak Performance Pty Ltd 2019

The right of Stan Rodski to be identified as the author of this work has been asserted by him in accordance with the *Copyright Amendment (Moral Rights) Act 2000*.

This work is copyright. Apart from any use as permitted under the *Copyright Act 1968*, no part may be reproduced, copied, scanned, stored in a retrieval system, recorded, or transmitted, in any form or by any means, without the prior written permission of the publisher.

HarperCollins*Publishers*
Level 13, 201 Elizabeth Street, Sydney NSW 2000, Australia
Unit D1, 63 Apollo Drive, Rosedale, Auckland 0632, New Zealand
A 53, Sector 57, Noida, UP, India
1 London Bridge Street, London SE1 9GF, United Kingdom
Bay Adelaide Centre, East Tower, 22 Adelaide Street West, 41st floor, Toronto, Ontario M5H 4E3, Canada
195 Broadway, New York NY 10007, USA

A catalogue record for this book is available from the National Library of Australia.

ISBN: 978 1 4607 5381 1 (paperback)
ISBN: 978 1 4607 0831 6 (ebook)

Cover design by Darren Holt, HarperCollins Design Studio
Cover image: Ball of yarn by shutterstock.com
Photograph on page 66 by Tim Gainey/Alamy Stock Photo; images on pages 46–47 and 67 and the repeated head icon and wool thread images by Shutterstock; all other photographs by Stan Rodski
Design and layout by Jude Rowe
Printed and bound in Australia by McPherson's Printing Group
The papers used by HarperCollins in the manufacture of this book are a natural, recyclable product made from wood grown in sustainable plantation forests. The fibre source and manufacturing processes meet recognised international environmental standards, and carry certification.

To Julie, for forty years of love and understanding

CONTENTS

INTRODUCTION	1
PART 1: WHAT IS MINDFULNESS?	**7**
Why Our World Needs Mindfulness	8
How You Can Achieve Mindfulness	19
Mindfulness and Energy Levels	34
PART 2: THE SCIENCE OF MINDFULNESS	**39**
'Do I Really Need to Read This Stuff?'	40
A Brief History of Brain Science	42
The Brain and Deep States	46
Consciousness and Memory	59
Balance and Connection in the Brain	67
PART 3: MINDFULNESS IN EVERYDAY LIFE	**73**
Core Skills	74
Mindfulness Activities	78
Energy Management 1: Physical Energy	96
Energy Management 2: Emotional Energy	100
Energy Management 3: Mental Energy	117
Energy Management 4: Mindful Energy	120
Energy Management Summary	124

PART 4: FROM MINDFULNESS TO WELLNESS: THE MIND–BODY CONNECTION   129

An Introduction to the Mind–Body Connection   130
MBC Research   135
Some Words of Caution   140

PART 5: HOW MBC WORKS   145

The 'M' Part of MBC: Mind   146
The 'B' Part of MBC: Body   149
The 'C' Part of MBC: Connection   154
Connecting the Dots: M + B + C = MBC   160

PART 6: USING MBC TO MANAGE YOUR HEALTH   167

Where to Start?   168
Believe in Yourself   172
Be a Rebel   176
Improve Your Emotional Intelligence   178
Personality Hardiness   182
Create a Compelling Future   191
Putting It All Together   207

AUTHOR'S NOTE   208

INTRODUCTION

In the course of my career I worked as a registered psychologist for over thirty years, mainly with executives and people dealing with high stress. During those years I became increasingly interested in the potential of the simple act of colouring in to help us rewire our brains. This would eventually lead me to develop my series of *Colourtation* books.

Tahlia came to my offices one day as a very stressed and anxious sixteen-year-old. She had severe agoraphobia – fear of open spaces – and found it immensely difficult even to go outside. She'd been suffering from the condition since she was about thirteen or fourteen, and it had worsened to the point where she'd been unable to attend school for the last two years. Her personal, social and family life was in tatters.

She had been medicated for her condition, but the more drugs she tried, the worse it got. Every time a health professional suggested something new, she only had to *think* about what she was being asked to do, and her fear would totally disable her. I might have been the seventeenth professional she'd seen in three years.

Her distraught parents had tried everything. This time they didn't want just another psychological intervention, but they'd heard about the work I'd been doing on rewiring the brain.

When they approached me, I said, 'Well, yes, I *will* take her on, and I want to trial a new technique that I've been working on.' And I asked, 'Does she like colouring in?' and they told me that she did. I said, 'Look, tell her to bring her favourite pencils with her, and I'll explain what she has to do.'

When I met Tahlia she told me that, like most of us, she'd really enjoyed colouring in as a very young child. In fact, she told me quietly that she *still* enjoyed it today.

I said, 'I'd like you to help me with an experiment.' (The word 'experiment' took some of the pressure off: whether it worked or didn't work was no big deal.)

And I said, 'I'm going to use an EEG – electroencephalography – machine. So we're going to hook up your earlobe, and we're going to look at the way your brain is communicating with your body, in particular the way it's communicating with a couple of your glands and organs that are producing this state of anxiety in you.'

What we were measuring were two neurotransmitters (chemical 'messengers'). One was the activation neurotransmitter, adrenaline, and the other was the relaxation neurotransmitter, dopamine.

First I just put her on the machine for five minutes while she sat there looking blankly at me, so I had a baseline from which to measure any improvement.

Then I said, 'OK, here we go. Here's a book of pictures. Now, they're not pictures of anything in particular, I know, they're just designs.' And I gave her some of the drawings I'd prepared for my *Colourtation* colouring books.

'Pick one,' I told her. 'Don't worry about me. Get your colouring pencils out, and you just colour. I'll tell you when to stop; it'll be after about five minutes. And we'll see what happens.'

While she coloured in, I encouraged her to practise mindfulness: 'Just forget about everything else and concentrate *only* on your colouring.' At the end of those five minutes, I told her to stop. I asked, 'So, do you feel a little more relaxed?'

She said, 'I do. I can't believe it.'

What really surprised her was looking at the EEG results before and after. The improvement was dramatic. She had relaxed, physically and mentally, and she could see it on the computer monitor. She felt better, and the computer reinforced and validated her feelings.

I didn't tell her which colours to use, but the colours she chose were dark – blacks, blues, purples and reds. This reflected her mental state; she felt under a lot of pressure, anxious about being depressed, and depressed about being anxious.

At our next meeting, Tahlia decided to use light and dark blue. Her EEG result further improved, as did her confidence in her ability to control her anxiety. By our third meeting, her colours were definitely brightening up, with the use of dark and light greens. She was starting to see that she could control the parts of her brain that affected her feelings and mood, and this realisation was reinforced by the technology that was monitoring her.

She had started the journey towards rewiring her brain.

*

From an art-therapy perspective, I suspected that the colouring idea would have benefits, but working with clients like Tahlia proved the science of it. This was the first time it had actually been demonstrated to Tahlia's *conscious* brain that she could control her *unconscious* brain.

After that, my discoveries around the healing power of colouring in and mindfulness grew more and more exciting. So a number of years ago I moved from working as a psychologist to focusing on neuroscience: an area I'd been involved in earlier in my career.

My journey had started in the world of mathematics; my first degree was in statistics. After that I did my doctorate in biostatistics, the use of statistics in health research, which was then a very new area in the neurosciences – so I didn't come at neurosciences as a brain expert, I came as an expert in measuring how the brain operates and what it does. And the reason I didn't pursue biostatistics at the time was that we didn't have the technology to prove what was happening in the brain in 'real time'.

Decades later, I've returned to the field as a cognitive neuroscientist. 'So, what does that mean, exactly?' I hear you ask. Well, neuroscience is the study of the way the brain operates. And *cognitive* neuroscience looks at the 'hardware' of the brain – how the cells and the neurons and the white matter and the grey matter and all that stuff that looks like firm pink jelly operates. I'm interested in why things happen – *why* part of the brain gets excited and tells a gland to secrete something that causes something else to happen. *Behavioural* neuroscientists focus more on the 'software' – the behavioural aspects that arise from that.

After all my years as a psychologist, one of the things I'm most interested in is stress. These days we're all stressed out. But if I were to summarise all of my learning over forty-odd years, I'd say that most people's stress starts with the complaint: *I don't have enough time.*

So this is a book for people who don't think they have enough time. There are only twenty-four hours in each day, into which most of us cram as much as we possibly can. Everything that doesn't get done in those twenty-four hours we start to get stressed about; other things that affect our lives we get stressed about. And eventually it has a major impact not just on our stress levels, but also on our general health.

Mindfulness is a technique that will help us quieten our very noisy brains, swirling with thousands of thoughts a day. This doesn't just automatically happen these days; we actually have to *tell* our brains that we want quiet. When you do that, your brain will try to change its thought patterns and processes. If you don't link your thoughts with a desired outcome, then anything could happen – in most cases *not* what you expected.

But perhaps the most important reason for pursuing mindfulness is so we can have a better health outcome, by improving our **mind–body connection.** The deeper we go into mindfulness, the more likely we are to enhance our brain, heart and immune-system functioning as well. The mind–body connection has big implications for a range of illnesses – immune-related diseases in particular, but also a growing list of other conditions (such as type 2 diabetes).

*

This is a book that says, 'You can get the benefits of mindfulness, of the mind–body connection, quickly and easily by applying a series of simple techniques.'

I've set out this book in a way that I hope will be easy for you to follow. It's divided into two halves. The first half **(Parts 1, 2 and 3)** looks at mindfulness and ways to bring mindfulness into your daily life, and the second half **(Parts 4, 5 and 6)** describes the long-term benefits of mindfulness that can be achieved through the mind–body connection. Within each half, I first introduce you to the topic, then I give you some of the technical information, then some practical advice – including quick and easy exercises you can use to get the most out of this exciting new science.

So go on – what are you waiting for? Read on to experience a new lease on life, as you discover how to eliminate harmful stress and heal your body …

PART 1

WHAT IS MINDFULNESS?

WHY OUR WORLD NEEDS MINDFULNESS

Why is it that while we are better these days at watching our weight, exercising and eating right and we have a better understanding of the lifestyle issues that can help us manage our stress, the fact of the matter is our health is getting worse? Lifestyle-related diseases like type 2 diabetes are growing, and our stress levels are getting out of control.

All of us are consumed by our high-tech, very fast world, surrounded by forces we don't really understand. But then this world actually starts to affect our health, and it becomes a serious, serious business.

People have often said to me, 'Isn't the world easier now that we have all this technology to help us?' In centuries past you would lose half your children as part of a normal existence; horrific wars would be a fact of life for everyone. Yes, those things *were* horrific, but they came and then they went.

Now we have wars every night in our living room – through our TVs, through our PCs, through our tablets, through our phones. Every day we hear and see and discuss the horror of kids being shot down in a school playground.

We live permanently surrounded by stressful events. We would never have heard about those things 100 years ago. We watch the six o'clock news and most of the first thirty minutes is just blood and murder and accidents. We sit in front of our screens and we think, 'Oh, well, I'm used to that.'

But you know what? There's a part of our systems that just keeps itself alert. Even if we're blocking those images out *consciously*, our subconscious system – which is one of the reasons why we're still on the planet – is telling us to be afraid, telling us that something is going to get us or hurt us or kill us. This fear is always just below our conscious surface. 'Look at that home invasion on the TV screen. When are *we* going to be next?' The likelihood of your being next is so small it's unbelievable, yet because it's in your face every night your stress hormones are constantly heightened.

Technology is delivering huge amounts of pressure in other ways too. Firstly, it's displacing us, taking over a lot of mundane jobs that once gave people an income. But it also means we can do everything more quickly, which has got everyone multitasking, working faster and faster, expected to achieve more.

The smallest error can really hurt you now! A one-second mistake in which you send an email to the wrong person can be a BIG error, depending on what the message says. We're all faced with little errors that have big impacts, which causes us a lot of stress too.

We're demanding ever higher performance from ourselves. But the usual method – putting in longer hours – has backfired. We're pushing ourselves harder and harder to keep up. Too many of us are reporting to our doctors that we feel we're at breaking point. We are getting exhausted, disengaged and sick.

Meet Brog

No matter how sophisticated and clever we think we are, our bodies are still genetically programmed to behave the same way as those of our prehistoric ancestors.

Imagine that a person called Brog lived 150,000 years ago. This was before towns and cities existed – long before those wars that wiped out whole populations – and the activities of Brog's daily life were largely survival-based.

Brog had to go out and hunt for his food, a dangerous venture that had to be performed regularly. Meat was important, as this protein was critical to the evolution of the human brain.

And nature ensured that Brog had the best possible physical mechanisms to ensure survival.

If Brog encountered a large and dangerous animal, such as a lion, he had a split second in which to decide whether to fight or run away. In other words, Brog perceived a situation involving either *challenge* – 'Aha! Fur-wrapped food with teeth and claws! Somebody get the fire stoked!' – or *danger* – 'Uh-oh! Trouble, run away!'

This is referred to as the **fight-or-flight response**.

For this to occur, a biological mechanism evolved called the **SAM system** (sympathetic adrenal medullary system). By means of the **sympathetic nervous system** (whose job is to perceive danger), the hypothalamus in the brain sends a signal into the adrenal glands (which are above the kidneys). The signal reaches a part of the glands called the **adrenal medulla**, which is responsible for secreting a hormone called **adrenaline** (epinephrine), which rapidly circulates throughout the body.

Remember when someone swerved in front of you on the highway without indicating? That sudden fright woke you up, didn't it? It probably felt like a combination of shock treatment and some seriously strong coffee. That was the effect of the hormone adrenaline. You just had a Brog fight-or-flight moment.

Adrenaline causes your body to stop digestion in the stomach so that all energy and blood can be redirected to the muscles. This will help you in either doing the 'macho' thing by fighting – in this case, swearing, hooting or extending a central digit to the offending driver – or mustering enough energy to get out of there.

Energy, in the form of glucose, is released for rapid action. Your heart rate increases, and the pupils of your eyes enlarge (dilate). Simultaneously, your sinuses and other mucous membranes stop secreting mucus. Your entire body is focused on one thing, and one thing only: either fighting, or running away as fast as possible. You literally feel wide awake when adrenaline is racing through your body. It's nature's caffeine.

After the challenge or danger is over, your brain switches from the sympathetic nervous system to the **parasympathetic nervous system** (responsible for the everyday working of your internal systems). The secretion of mucus resumes, your heart rate and breathing slow down, and your digestion kicks back into operation. (We'll look more at the sympathetic and parasympathetic nervous systems in **Part 5**.)

(Interestingly enough, sexual arousal, orgasm and post-orgasmic 'glow' involve the same process, switching from the sympathetic to the parasympathetic nervous system. There's a direct biological connection between sexual arousal and the physical arousal related to danger, and it's not surprising that some people confuse the two.)

External versus internal threat

The SAM system was originally designed to deal with external threats to survival in a harsh and dangerous world. There were no supermarkets where Brog could buy food, so he had to hunt to survive. It was also extremely unlikely that he would remain free of injuries for long – whether major injuries from fights with animals and other humans, or simple cuts and scratches from living a primitive life. The challenges and dangers in his life were short-term and external. The SAM system was ideal under these circumstances.

But for the most part, today's world is a very different place. How many people do you know who – of necessity, not choice – encounter wild and dangerous animals on a daily basis? How many still hunt wild animals for food? (And no, an irritable cow does not qualify as a wild animal.) Also, injuries are generally much scarcer than they used to be.

We have grouped together into large communities, an arrangement that offers protection for each community member, and we have eliminated most of the external threats from our environment. Our food supplies are regulated to such an extent that, if we have the money, we can basically buy anything we want (though getting a job to *earn* the money can be another issue).

There are still certain external dangers, including crime, but there are systems in place to regulate these. Many people also have some degree of choice regarding these things, such as moving to a different neighbourhood or installing a security system. There are certainly parts of the world where many dangers are still external, but even these are much less extreme than they were for Brog 150,000 years ago. For most people nowadays, threat is no longer external. Instead, most of the threats we face are internal. These include fears and anxieties around failure and rejection.

But surely we know the difference between real (external) danger and imaginary (internally perceived) danger? Why would we respond to a fear of failure as if we were facing a hungry and dangerous tiger? How is that possible?

There are two answers to this question:

THE NEUROSCIENCE OF MINDFULNESS

1. Your **conscious mind** may know the difference, but your **unconscious mind** may not. The unconscious mind does not distinguish between real situations and imaginary ones. Have you ever had a nightmare and woken up dripping with sweat, your heart pounding? You were safe in bed, yet your body responded to the images in the nightmare as if they were real. Similarly, for the unconscious mind, 'If he leaves me I will just die!' is as real as if you were actually facing a life-and-death threat outside of yourself. The other important thing to remember is that your body responds to your unconscious mind *more* than it does to your conscious mind. (We'll learn more about this in **Part 2**.)

2. There is plenty of evidence to demonstrate the phenomenon called **behavioural conditioning**. The first person to document and research this was Ivan Pavlov, in 1928. He developed what he called a 'conditioned response' in dogs by ringing a bell when he fed them. After doing this for a while, he could get the dogs to salivate as if there were food on the way, just by ringing the bell. The dogs became conditioned to associate the food with the sound of the bell. This association continued, even when the bell was rung *without* the provision of food. The point is that, since the time of Brog, our prototype cave person, we have been genetically conditioned to respond to danger in a specific way. Actual injury or pain has been linked to failure, and this association has been carried forward to today, despite the absence of *real* danger if you fail. A long time ago, if you failed, you could die. Today, this is unlikely, but it still *feels* as if it could happen. The same could apply to a whole range of emotional perceptions. We now fear emotional hurt in exactly the same way as we once feared actual hurt.

Short-term versus long-term stress

Ironically, such internal threats and anxieties (beliefs, perceptions) are more difficult to deal with than an angry and hungry lion.

When you face a lion, you quickly discover whether you can fight and win, or run and survive. If neither option works, you won't notice, because you'll be dead! Regardless, the danger is resolved one way or the other in a very short space of time.

Normally, the surge in adrenaline has only a temporary boosting effect on the immune system. The research clearly indicates that short-term events – such as watching a scary movie or experiencing a sudden fright – cause a spike in our adrenaline levels, but this spike quickly dissipates and we return to normal within fifteen minutes to half an hour.

However, fear of rejection or failure can last for months, even years. The 'danger' is chronic and long-term. And this is the key to the problem: **your body is not designed to handle long-term threats and dangers**. If your body is stimulated to produce adrenaline for too long, it causes the SAM mechanism to backfire. The SAM system works wonderfully for short-term events and circumstances but becomes destructive when the perception of 'danger' becomes chronic.

A perfect example is the need in our society to be the best, to win and to avoid failure at all costs. How many movies or television shows feature people climbing to the top of their profession by working long hours, day in and day out, and ignoring family and relaxation? These people are admired and selected for promotion. No one seems to notice that these people also don't last very long in that profession, because they simply burn out, get sick with a chronic illness, or have a heart attack.

Think about the average high-powered business executive, living with constant challenge, which leads to the over-activation of the SAM system and continual wear and tear on the heart. Generally speaking, the need to avoid failure, and to win at all costs, is called the **stressed power syndrome**. (We'll encounter this again in **Part 6**.)

Major life events – marriage, divorce, death of a spouse, moving to a new town, starting a new job, sitting exams – can have a similar effect. (**Part 5** includes a longer list of life circumstances and psychological factors that have been shown to weaken the immune system.)

But a lot depends on how we *perceive* these events. Here are a couple of examples:

1. If you divorce your spouse, and afterwards you view the failed marriage as a simple mistake that does not reflect on your overall self-worth, then the effects will probably be relatively short-term. However, if you perceive your divorce as a reflection of some fundamental failure, or as indicating there is something wrong with you, or if you worry about never finding another person to love you, the effects will be different.

2. If you feel unhappy in your job or marriage, and do something to change your circumstances, the unhappiness will have a comparatively short-term effect. But if you perceive that you're 'trapped' in the job or marriage and can't leave, you will probably experience feelings of helplessness and hopelessness, resulting in long-term, chronic stress.

Remember that stress is not caused by an event – it is caused by your perceptions or beliefs or decisions about that event.

There is no such thing as a stressful event *per se*, because one person may experience it as positive and challenging while another may experience it as devastating. An extreme case is the death of a loved one. If the death is sudden, then the sense of loss will be tremendous. However, if the person died after a long illness, or at an advanced age, then you may feel a sense of peace, even relief, both for yourself and for the deceased.

What happens to your body when it's stressed?

When the SAM system is activated by a threat – or a *perceived threat* – it releases chemicals called neurotransmitters – in particular, as we've seen, the neurotransmitter (hormone) **adrenaline**. These neurotransmitters travel along pathways, taking part in a game of 'kiss and tell': when one adrenaline neuron (nerve cell) finds another adrenaline neuron, they will kiss, and tell each other, 'I like you', and this triggers a hormone release.

That hormone is **cortisol**.

Once you've dealt with the threat – you've fought or run away and survived, or you've realised it was a false alarm – your system starts to settle, and your parasympathetic nervous system takes over once again. It discharges another neurotransmitter, called **dopamine**, to relax your system. The dopamine generally does that by releasing **serotonin**, which then negates the cortisol in the system.

What scientists have only discovered in recent years is that cortisol needs to be quickly dissipated. And what they've found is that we live in a world in which, for many of us, the cortisol stays in our system constantly. Long after you have been 'stressed', your body keeps pumping out adrenaline.

Think of it as a little bit like bathwater. If you drain the water out quickly, nothing happens. But if you let the water sit there, when you eventually do remove it, you leave a scum mark behind on the bath.

That's actually exactly what happens when the cortisol stays in the brain too long – which for many of us occurs frequently; many of us are on edge all day. The brain's equivalent of the scum you might find on the edge of the bath is called **amyloid protein**. And that protein literally builds up in blood vessels, in both the brain and the heart, leading to serious health consequences – the extent of which we're only just beginning to discover. (See the discussion of the mind–body connection in **Part 4**.)

Here's a graphic that shows what happens when the SAM system is activated, why it happens and the effects that prolonged activation can have on your body.

Reaction	Purpose	Long-term effects
Heart rate increases	To pump blood faster	High blood pressure
Breathing quickens	To increase blood oxygen level	Chest pains from tired diaphragm muscles
Digestion stops	To divert blood from gut to muscles	Ulcers
Blood leaves head, hands and feet	To divert blood to muscles	Cold hands and feet, migraine headaches
Coagulation of the blood increases	To minimise blood loss if body is wounded	Blood clots, strokes, heart attacks
Extra sugars, and insulin, which breaks these sugars down, are released into the bloodstream	To rapidly improve the body's energy levels	Low blood-sugar levels, tiredness
Stress messages are sent to muscles	To prepare muscles for action	Muscle fatigue

All sorts of symptoms can indicate that you are overstressed, but here are five sure signs:

1. Increase in physical problems and illness (see **Parts 4**, **5** and **6** for more on this)
2. More problems than usual with relationships
3. Increase in negative thoughts and feelings
4. Significant increase in bad habits
5. Exhaustion

So, what do you do? Thank Mother Nature for the warning signs and make an agreement with yourself that something has to change. Identify what is causing you the stress and begin to search out your options to resolve it. As we'll learn, one key is to focus on changing your lifestyle by managing your energy, not your time. (Obviously if you have issues of deep trauma in your past you will need more rigorous intervention than this book can provide.)

But you can also take steps right now to take control of your stress levels, both in the acute short term and in the chronic long term. And you can do it almost instantaneously. With mindfulness.

HOW YOU CAN ACHIEVE MINDFULNESS

Mindfulness can be thought of as like a rest in music.

A rest is a place where the music stops for a brief moment, in keeping with the overall rhythm of the song. It's an **interval of silence**. A rest is just as important to a song as the music itself. It informs the music, just as mindfulness informs life.

Mindfulness can be thought of as a rest from the busy activity of our brain. It puts a brief hold on the chatter.

Each one of us has in excess of 60,000 thoughts – and the emotions they evoke – swirling through our brain every day. During this process we rarely notice, and even more rarely bring our attention back to, the task we're meant to be engaged in. The average attention span is now eight or nine seconds.

It's easy to understand how the brain, overwhelmed by this constant flow of thoughts, feelings and sensations, can get confused and unfocused. When your mind is cloudy and cluttered, you may *think* you're working effectively, but you may be wrong.

Think of the brain as a snow globe. When we shake the globe, it's just like what happens in our brain when we're worried, angry, distracted, stressed out or revved up, and all our thoughts and feelings are whirling around.

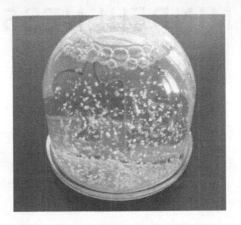

If we continue to watch the globe, and no more shaking occurs, the snow settles to the bottom.

This is what mindfulness does to your brain. It clears your thoughts and feelings and helps you feel more relaxed and better able to concentrate.

In physical terms, what it does is take the cortisol out of the system and encourage the release of dopamine and the resultant serotonin. They go out of the system when we're stressed – naturally, as why would you want to contemplate the mysteries of the universe as a bus is about to hit you? Your brain is too busy deciding whether you should jump, move or just stand there and die.

Mindfulness encourages the opposite of a fight-or-flight response: deep relaxation. It will also help you:

- Recharge your batteries
- Help you keep things in perspective
- Encourage normal emotional and physical healing
- Give you a sense of control
- Help you avoid taking out your stress on those around you
- Give you resilience so you can bounce back from stress
- Enhance creativity and concentration

And how long does it take to restore your system so that it has a healthy balance of cortisol and dopamine? About a millisecond. That's all your system needs to shock you out of that cortisol-fuelled, sitting-on-the-edge-of-the-seat, gritting-your-teeth world – the world many of us inhabit all day.

What is mindfulness?

So what is mindfulness, exactly?

Once upon a time we used words like relaxation, hypnosis, meditation, yoga ... Well, mindfulness is a relatively new term that's plucked out of what is the essential aspect of all of them, which is being aware of where you are in the moment, and being able to focus on the activity that you're doing, to the exclusion of everything else.

The benefits of meditation are now generally accepted and understood. It's a clear winner in helping people relax. More than nineteen replicated studies have proven that practising meditation reduces hypertension.

But mindfulness is not about meditation. In mindfulness, you do not need to sit still with no thoughts for twenty to thirty minutes. You do, however, need to learn the process of gradually becoming better at dismissing distracting thoughts and gaining the ability to *focus*.

Mindfulness involves **paying attention to something, in a particular way, on purpose, in the present moment, non-judgmentally.**

Let's look at this a little more closely.

Paying attention to something ... This could be anything you choose to pay attention to. It often begins with paying attention to your breathing (see **Part 3**, page 101), but it could also be paying attention to one of the mindfulness activities described later in this chapter.

In a particular way ... This is about *focused attention* on the task you've chosen. You need to think *only* about what you're doing, not about anything peripheral. It's not an opportunity to start thinking about bills that need paying, or reading the text that just came in from the bank about your mortgage. And it's not your 'general thinking time'; it's your blank thinking time.

On purpose ... This is about making a conscious decision to pay attention to the task you've chosen. Your intention, for the next few minutes at least (longer if you wish), is to be totally absorbed in what you're now doing. You actually have to *tell* your brain that is your intention. It won't know what you're doing, and it will prioritise its activities based on your emotions or your mental energy or your physical needs. If you don't tell your brain, it will be thinking about what you have to do next, or a million other things, even the meaning of life, and you will be off track within a minute. Paying attention uses up an immense amount of the brain's energy, but if you actually combine that with relaxing and focusing on what you're doing, the results will be highly beneficial.

In the present moment ... This means dismissing all thoughts of the past or future that may arise in the present. Tell yourself: 'My intent is to do this but to remove everything else from my mind and relax.'

Non-judgmentally ... Don't judge or be critical of yourself while paying attention to your task, don't put pressure on yourself to do the task well, and don't compare yourself with others. A key feature of mindfulness is that it is non-competitive, and free of deadlines and expectations. It's about the **process**, not the **outcome**.

All sorts of everyday activities can help you achieve mindfulness. The important thing is to do something that will help you switch off, relax and flush the cortisol out of your system.

Many people carry out mindfulness by doing what they call a 'physical body scan' (see page 81). They progressively move up the body from the tips of their toes to the top of their head and try to sense where the pressure and tension are, so that they can release them. The whole body enters a state of deep relaxation, or mindfulness.

People of many faith backgrounds may use patterned and repetitive forms of prayer to achieve a state of mindfulness; being in a religious space such as a church also helps create a distance from the demands of the outside world. Other people use a range of deep-breathing techniques.

One of the reasons why we find it so hard to relax is that technology has made so many traditional tasks redundant

– focused, repetitive tasks that naturally encouraged mindfulness. Instead of gardening or knitting, many of us now employ gardeners, and buy scarves or jumpers. People even pay to have their dogs walked. For many people, mindfulness involves rediscovering some of those 'slower' activities that come from an era when people had more time and were less stressed as a result. Many of these activities are making a comeback in our world today.

Mindfulness activities can be skill-based – crafts and other hobbies – or semi-skill-based, like drawing, but some require no skills at all. Even when you're walking, having breakfast, washing the dishes or entangled with your lover, you can still be participating in mindfulness. As we've seen, mindfulness simply means paying attention to what you're doing while you're doing it. As soon as you notice your mind has wandered, you need to return your attention as quickly as possible to the task at hand. The activity also has to involve the three elements of **repetition, pattern** and **control**. For instance, **brushing your teeth** could be a way of practising mindfulness. You brush one way, you **repeat** it the other way, and you create a pattern in the way you clean your teeth. But you do it within a boundary: you don't go outside the mouth, you stay inside the mouth. Go outside and you've broken the pattern, removed the **control**, and, all of a sudden, you're upset – because you've got toothpaste all over your face!

You could also practise mindfulness **under the shower**. Surrounded by the tactile sensation of hot water, you could easily daydream yourself off into a million different scenarios. But if you stand in the shower and intentionally concentrate on feeling the water hit you, sense it going down your arms and legs and back, sense the change in the sensation as it gets colder, the relaxation benefits could be tremendous. Your intention will be to be aware of the water and you in the shower. (Actually, baths are my number-one recommendation for children who have nightmares. Give them some malted milk and put them in a hot bath, because it actually stimulates a relaxation response, and their night terrors will be gone – guaranteed.)

The key to making sure these activities have a positive impact on the brain is to **focus on the process and not the outcome**. Outcome focus – 'How good is my drawing/embroidery/woodwork?' – sets the brain up for **competition**. Competition, even with yourself, immediately unravels all the good chemical secretions that the brain was previously producing. Your brain will now be engaged in very limited activity, in only one direction: winning.

So, pick an activity you love, don't put pressure on yourself, and don't worry about the results. Knitting is a hobby many people enjoy, but for others, knitting is *not* a positive experience, because they still have drawers filled with scarves that didn't end up right. Other people love jigsaws, but they give themselves a timeline and get stressed over finishing, which is totally unproductive. But losing themselves in the activity – working out where the pieces go, fitting them in and putting the picture together – could be a rewarding mindfulness experience.

You'll find more guidance on how to practise popular mindfulness activities in **Part 3**.

The Colourtation method

Not everyone, however, has the time, the opportunity or the headspace to practise such activities on a daily basis, especially if they are out at work or in school – which is why I developed the *Colourtation* method.

Mindfulness in the office

During my thirty years as a practising psychologist, I would continually see busy executives who were becoming more and more stressed. In my normal clinical way, I would say to them, 'You need to change jobs, but of course you can't do that, because you've got golden handcuffs on you, you've got a family and a mortgage, and so you're caught in the treadmill.' So I would tell them, 'Look, go off and relax – do some running, or do some breathing or meditation.'

But what I found was that many of the executives who tried breathing or meditation would come back to me soon after and say, 'Oh, Stan, I don't know whether it works. I sat there and I thought about a problem I've got at work, and that's all I did for the fifteen minutes, or half an hour, or hour. I sat there, I couldn't stay on the mantra, I couldn't concentrate, I just couldn't do it.' Other common responses were 'I kept forgetting to do it.' 'When I tried it, I fell asleep.' 'I felt silly doing it, and it's boring.' And above all: 'I DON'T HAVE TIME!'

The fact is that most people know about meditation, they know about yoga, they know about mindfulness, but they see them as time-consuming activities, which they therefore never do, because guess what? Your brain's managing your energy, and if it spots something that's going to cause you more grief – more lost time, because it takes you half an hour to get to your meditation class, and half an hour to do your meditation class, then half an hour to get back to wherever you were – at a subconscious level, it won't allow that. No, it just changes the priorities in your subconscious system, and the meditation class never happens.

I could have just been another voice saying, 'Stop looking at your phone, stop going to multiple meetings, stop being on the internet all the time, stop looking at your computer just before you go to bed', behaving like the proverbial nag. But I realised there had to be another way that would work for people living in the digital world. There had to be a way around the issues I was seeing in my practice every single day. There had to be a way of removing the cortisol from the brain *quickly*.

Mindfulness in schools

As well as working with executives, I was doing a lot of work in schools, with kids as young as infant/prep-school age. You only have to talk to teachers to understand how these kids can get really stressed – and it's not necessarily because they have mental issues or there's something major going on in their lives, it's just that the pressures of our world have a compounding effect. We're all under pressure – it keeps us going in the right direction – but of course there's stress and there's distress. Some of these kids are highly strung, while some of them don't get too fussed about anything, but all of them are somewhere on that spectrum. As they get a bit older, of course, hormonal factors and other things also enter the mix – and it all affects learning.

In years gone by teachers would ignore these issues, and just tell little Johnny or Mary to be quiet and calm down. But what the schools have now realised is that if they can help the kids to control their stress early, and teach them to relax, they can actually improve the outcome for a lot more students. This has become a big movement worldwide: for instance, the American neuroscientist and psychologist Martin Seligman has done a whole lot of work in this area. It's helping to create a larger pool of kids in the 'normal' range.

Even in our modern world of smartphones, tablets and computers, colouring in continues to be used to teach children hand–eye coordination, in preparation for learning to write. They start colouring, and they go all over the page, but they're getting used to the hand movements. And then the teacher will say, 'Now, look, try colouring in this big area here, but try to stay there', which is very difficult for children, because their brains have yet to work out how to coordinate hand and eye movements. But essentially they start to colour in smaller and smaller areas within the lines, and as soon as they start doing that, their hand–eye coordination will be at a level where they're ready to learn to write.

The process of writing is an immensely complex task for the brain: not only does it require coordination, but it also involves memory, it needs to be done in a certain way – usually, in our culture, left to right – for older kids it means connecting the letters, and it also involves comprehending what they're doing. You only have to be a parent of a child who has dyslexia or dysphasia, or one of the many other language disorders – which are generally childhood conditions, because adults tend to hide them – to understand how much tasks like writing demand of the brain.

It's always intriguing to watch little children (Kindergarten to Year Six) when they're asked to do colouring in, and the effect it has on them behaviourally. The thing that I observed when I began working with young children was that when the teacher announced, 'It's colouring time', the mood of the room changed. Suddenly the kids were all focused, engrossed in the task of picking their colours then colouring between the lines. You ask any primary-school teacher how to settle a class down, and they'll say, 'Get the colouring books out.'

Developing Colourtation

I'd been starting to look at art therapy – healing using creativity – as a way to achieve mindfulness. I sensed there was something in the process that would help my stressed clients. Even though mindfulness is a mental thing, for most of us, trying to connect it with a physical activity is important, because we're very visual beings. Our eyesight is an incredibly important part of our lives, and the visual cortex in our brain is closely connected with our movements and our speech. Colour is important too, because its soothing effects, through its impact on our glandular activity, are quite amazing.

Most adults remember colouring fondly from their childhood. So I thought, 'Well, if I could actually achieve the same thing with adults that happens with children …' I thought that if my clients did something that was associated with a time and place in which they'd felt happy and secure, the brain would recognise this and immediately open the neuronal pathways it associated with those feelings.

Even though colouring has always been a way to aid the process of learning to write, the neuroscience said it was more complex than that, because it in fact enabled the brain to access memory, coordination, comprehension – all the things that are required to enable you to put your thoughts to paper. And it sparked up in the brain the relaxation that it had been missing.

Next I thought about the *kinds* of pictures I would ask my clients to colour. I knew I couldn't give my executives a picture of, say, Donald Duck – something recognisable, where there would be pressure to colour it a certain way – nor a picture that they couldn't finish quickly. (Being over-achievers, they had to finish *everything*!)

So I enlisted the help of my soon-to-be son-in-law, Jack, who was studying architecture. I asked him, 'Can you help me with some images?' And I gave him the parameters of what I needed.

I knew that the brain is most comfortable when it experiences:

1. Pattern

2. Repetition

3. Control

The brain needs to create **patterns**, and these patterns are created through **repetition**, through habitual activities that are done within a **control** or boundary – colouring within the lines, for example. When the brain has all of those three, it relaxes. The brain will be even more relaxed when these activities allow **creativity** in a **non-competitive** way. This became the core of the *Colourtation* method.

So I said to Jack, 'Look, the images I need could be any sort of line drawings, but they've got to be able to create geometric patterns, when you add colour with repetition.' And so he produced some images for me using architecture software, and I started experimenting with them and using them with my adult clients.

After they'd coloured in for five minutes, I wouldn't even need to ask them, 'Do you feel better?' I would just look at their results on my computer screen and turn it around to them and say, 'You know how we've never been able to get you relaxed without half an hour of breathing techniques – which you never do after you leave? Well, if you do the colouring instead, look at what happens to your system. You immediately ease your tension physically, and increase your relaxation mentally. And that's after five minutes! And I know that in a nanosecond you will remove the horrible cortisol out of your brain that causes all the damage.'

And that was where it all started. So then I took my pictures into the schools – all in trial mode, testing, testing, testing. One girls' school allowed me to do a study with a class of Year Tens – around the age of fifteen. They typically had me in for the period after morning recess. You can probably imagine: the girls had just had a break, they were in full flight, and now they were in for their next class, and they were so noisy. It's hard to believe just how rowdy twenty-five fifteen-year-old females can get.

The class teacher would announce: 'Girls, here is Dr Rodski, we're doing a bit of research with him. On your desks are colouring books and pencils.' And the students had been supplied with our *Colourtation* colouring books.

The teacher instructed them, 'Please pick a drawing, and colour in until I tell you to stop.' Within seconds, you could hear a pin drop. Despite all the things the girls had just been doing and thinking about, the whole classroom immediately went quiet.

I did the same thing at another school with a class of Year Eight boys: an even more boisterous lot. I would demonstrate the science to the students afterwards on the computer. I'd pick one student and show them what had happened to their brain, explaining how the colouring in had settled them down. It was fantastic evidence of just how powerful the outcomes of *Colourtation* could be – and they only improved in the older students I worked with.

Finally Jack and I published our series of *Colourtation* books. To our surprise, they were a worldwide success, and were even featured on Oprah Winfrey's 2016 Christmas Wish List. We were ahead of the game, but since then there have been many imitators, not all of them featuring the same scientific backing. Still, it's exciting to know that so many people around the world are now using adult colouring-in books to give their brain the things it craves.

Using *Colourtation*

I still speak at hundreds of schools every year. I encourage the kids not just to see colouring in as some sort of creative pursuit, but also to regard it as a way to help keep themselves calm.

I'm finding the kids now do a lot more things like meditation in their classrooms. A number will get value from just sitting on a mat, or sitting at their desk, and closing their eyes and stopping for five minutes. But just like adults, a lot of children find it very hard to stay still. There's a reasonably large percentage that can't quite seem to do meditation. And so the colouring gives them something to do rather than just sit, close their eyes and try to remove all thoughts from their head. Instead, they can simply be told: 'The only thing you should think about is your colouring.'

Over the years I've been able to follow many of these groups of children, and I've consistently found that those who take part in meditation, or mindfulness in the form of colouring in, obtain *significantly* better school results.

The University of New Mexico in the United States is now using the colouring books with their maths students, and encouraging them to do some colouring in before sitting their exams. They tell the students, 'You'll end up with around five extra marks if you colour in beforehand.' The students' autonomic nervous systems are in a far more stable place before they see the first question, and they move through the exam paper much more quickly.

Find your own way

Nevertheless, about 2 per cent of the population find colouring a stressful activity. Instead of having had a pleasant experience with colouring as a child, they've had a very *un*pleasant experience. Perhaps they were made to use their right hand when they should have used the left one. Perhaps they were unable to keep between the lines and therefore became self-deprecatory, or perhaps they had a terrible teacher who told them, 'You can't do it right.' And then there are the very, very few people who just abhor the act of colouring in.

In fact, not everyone even finds 'relaxation' relaxing. In one study of anxious patients, nearly half reported feeling more anxious when they started to meditate. When they attempted deep muscle relaxation, nearly a third suffered increased restlessness, sweating, heart pounding and rapid breathing. Maybe these patients were simply reacting to the new sensations they were experiencing; maybe they felt fearful and no longer in control; or maybe they had an aversion to the self-observation involved.

The point is that **mindfulness is *not* one-size-fits-all**. This is why wrapping the principles of mindfulness around activities that make sense for *you* is so important. And if you can really get your mind around the process, you can apply it to any number of things you do on a regular basis, or even every day.

> *'I've consistently found that children who take part in meditation, or mindfulness in the form of colouring in, obtain significantly better school results.'*

MINDFULNESS AND ENERGY LEVELS

Practising activities like those we have just discussed is really just the *first step* in using mindfulness to reduce stress and improve health – it does a great job of reducing the effects of daily stress. But to *prevent* a lot of the stress from occurring in the first place, you need to start applying mindfulness to your daily routine, and the way to do that is to learn how to manage your energy.

Our brain has the extraordinary skill of being able to produce electrical activity all the time. Whether we are awake or asleep, it is continuously receiving and processing information. It is trying to make sure we are able to do the things we *consciously* want to do, while continuing all the things it *subconsciously* has to do – keep our heart beating, keep our digestion going, keep all our secretions going, keep our blood vessels moving along, moving the oxygen in, and the carbon dioxide out. Even when our conscious system shuts down, our subconscious system will continue doing these things automatically.

We have three pounds of firm pink jelly sitting on our neck, being forced down by gravity all day long, while our brain is trying to maintain our energy between the time when we wake up and the time when we finally become unconscious again. No wonder our brains – and bodies – get exhausted!

Do you know any of those people who seem to be just as busy as you at work, and have countless things they do *outside* of work, and seem to fit in a whole lot of stuff on weekends, and always manage to go on long holidays, with no problem at all? Do you look at yourself and ask, 'How do they do that?'

As I wrote in the **Introduction**, the heart of all the stressors that affect your health is time. But the answer isn't **time management**, it's **energy management**.

I can hear you saying, 'Oh, well, that's good for *them*, they've got the energy, I haven't.'

No, no, no.

Your time may be a limited resource, but the great news is that your energy is renewable. Energy can be brought into your system as easily and effectively as into anyone else's. You just have to learn how to do it – through mindfulness.

If you have no time to slow down, you need to create time. In five minutes or less, once or twice a day, you can rid your system of harmful amyloid protein and return to zero. It might just be a matter of spending a few minutes of your time colouring in at your desk, so you can work more productively for hours afterwards. Or it could involve taking just a few minutes to go and dig in the garden, understanding that these few minutes are going to give you the energy to go right back into the house and fix three things that you've been procrastinating about for the last two weeks.

But to recharge yourself, you also need to recognise the cost of your **energy-depleting behaviours** and take responsibility for changing them.

One of my clients, Bob, was a highly respected thirty-seven-year-old partner at a major accountancy firm, married with four young children. He was working twelve- to fourteen-hour days, was perpetually exhausted, and found it difficult to fully engage with his family in the evenings, which left him feeling guilty and dissatisfied. He slept poorly, made no time for exercise, and seldom ate healthy meals, instead often grabbing a bite to eat on the run or while working at his desk.

Bob's experience is far from uncommon. Many of us respond to rising demands in the workplace by putting in longer hours, which inevitably take a toll on us physically, mentally and emotionally. That leads to declining levels of engagement, increasing levels of distraction, and for employers, high turnover rates and soaring medical costs among employees.

Most employers invest in developing their employees' skills, knowledge and competence. Very few are interested in helping build and sustain their employees' capacity – their energy – which is typically taken for granted. In fact, greater capacity makes it possible to get more done in less time, at a higher level of engagement and with more sustainability.

The rituals and behaviours Bob established, through working with me to better manage his energy, transformed his life. He set an earlier bedtime and gave up drinking, which had disrupted his sleep. As a consequence, when he woke up he felt more rested and more motivated to exercise, which he now does almost every morning. In less than two months he lost 5 kilos. After working out he now sits down with his family for breakfast. He still puts in long hours on the job, but he renews himself regularly on the way. He leaves his desk for lunch and usually takes a morning and an afternoon walk outside. When he arrives home in the evening, he's more relaxed and better able to connect with his wife and children.

Maybe your work situation is nothing like Bob's. Maybe you're a student, maybe you do unpaid work, maybe you're unable to work or retired. But the same principles can be applied to *any* life situations that occupy large amounts of our time. Whatever circumstances cause us stress and exhaustion, energy management is the answer.

Energy falls into four main categories:

1. Physical energy

2. Emotional energy

3. Mental energy

4. Mindful energy

To manage your energy, not your time, you'll need to focus on *all* these areas. I'll show you how to do that in **Part 3**. But first, in **Part 2**, I'm going to explain more about *why* mindfulness has such a positive effect on your brain.

PART 2

THE SCIENCE OF MINDFULNESS

'DO I REALLY NEED TO READ THIS STUFF?'

When you read the word science, it may make you want to put this book down and run for cover. If that's the case, take a deep breath and relax ... You do not have to understand any of the in-depth scientific information in order for the methods in this book to work. In fact, if scientific detail scares you, go ahead and skip the 'technical' parts of the book – **Parts 2** and **5** – altogether. Not understanding how these things work is not going to stop you from actually doing the processes I've included later in the book.

Then why would you need to read the scientific stuff? If you want to see whether mindfulness and the mind–body connection can offer you some assistance in managing your health and wellbeing, some knowledge of how your body works can go a long way in helping you decide what will work in your own situation, and what is less likely to work. Though intellectual knowledge can only lead you to the door – the actual healing lies *beyond* the door, in the realm of feelings and emotions – it can also provide some impressive conversation topics at parties!

But seriously ... without some basic understanding of how mindfulness works, this book would be doing nothing more than asking you to trust the opinion of a stranger – me – who is telling you, 'Do this, do that, and don't ask me why – I'm the expert!' So why not give the technical information a go ... and don't stress about not understanding everything.

The purpose of this material is to give the average reader a straightforward, but logical, summary of how our understanding of mindfulness and the mind–body connection (MBC) is supported by clear scientific evidence, and is not just a matter of sheer belief. The summary is, however, purposely over-generalised. So I ask expert readers – neuroscientists, biochemists, psychologists, medical doctors – to take a deep breath and consider who this information is aimed at. My objective is to make the neuroscience behind mindfulness and MBC accessible to the public, and to this end, I have consciously sacrificed theoretical and technical detail by simplifying complicated concepts. If you are a medical or scientific professional, please forgive the generalisations. Read between the lines, and add the detail to your satisfaction.

To kick things off, it will be helpful to learn a little bit about how the science of the brain, and the principles of mindfulness and the mind–body connection, emerged.

'You don't have to understand any of the in-depth science for mindfulness to work.'

A BRIEF HISTORY OF BRAIN SCIENCE

In the middle of the fifth century BCE, ancient Greece had three outstanding centres of medical science. The oldest of them was in Croton (modern Crotone), a Greek colony in what is now Calabria in southern Italy. Alcmaeon, Croton's foremost physician, researcher and lecturer, was the first to write about the brain as the site of sensation and cognition. Because he was a practising physician, his approach was entirely clinical, developed through the study of brain-injured patients.

About 500 years later, Claudius Galenus (129–199 CE), more commonly known as Galen of Pergamon (now Bergama in Turkey), used piglets to perform the first recorded experiments on the brain. As perhaps the premier medical researcher of the Roman period, he devised a number of experiments to demonstrate that the brain controls all the muscles in the body through innervation (stimulation) by the cranial (brain) nerves and the peripheral nervous system (nerves outside the brain and spinal cord).

Ever since, this organ with the consistency of a soft-boiled egg, floating in spinal fluid, has continued to challenge medical researchers. From anatomy to physiology and, much more recently, from neurochemical reactions to electromagnetic fields, the brain has slowly been yielding its secrets.

Psychosomatics

People believed that emotions, beliefs and attitudes could affect physical health long before there was actual scientific proof. More than 2000 years ago Aristotle stated that there is a connection between melancholy (depression) and cancer, but it wasn't until the early twentieth century that the field of psychosomatics (from the Greek words *psyche*, mind, and *soma*, body) developed. However, because there was no real way of *proving* that thoughts or emotions could either cause or cure illness, this field of research did not carry much weight within the wider medical profession.

This scepticism was and still is justified, because even if you can prove that there is a connection between, for example, cancer and depression, this does not mean that the depression *caused* the cancer. You could logically argue that people with cancer are more depressed than people without cancer, because cancer is frightening and painful, and the person naturally has negative thoughts about chemotherapy, surgery and possible death. It could also be argued that something about the cancer cells causes the release of chemicals that lead to depression.

The point is this: psychosomatics, in a general sense, could not prove anything more than the fact that certain diseases are *associated with* certain thoughts and emotions.

Fortunately, thanks to new technology and research into the mind–body connection, these associations can now be tested to see which comes first: the illness, or the thought/emotion.

The emergence of MBC

Beginning in the late 1970s, a new approach to the mind–body dynamic slowly developed into what we now call the mind–body connection (MBC). This has now replaced the field of psychosomatics.

Frustrated with the 'soft' approach to mind–body research up to that point, a number of researchers – mostly in the medical and psychiatric fields – began to investigate the biological mechanics of how, if at all, psychological states impact on the body. Instead of using disease as the measurement, they began to look at the body's *immune system*, and all the various cells and systems that determine when, how and if disease forms, and how physical healing occurs.

This was initially difficult. For example, the technology and equipment necessary to count the numbers of a specific cell in a millilitre of blood were slow and very expensive.

By the mid-1980s, this had all changed dramatically, largely due to intensive research into cancer, then human immunodeficiency virus (HIV) infection and acquired immune deficiency syndrome (AIDS). Millions of dollars began to be invested in the area of immunology. A direct outcome of this unprecedented funding was the development of equipment to rapidly – and relatively cheaply – test and measure the finer details of what occurs in the body from moment to moment. An example is the flow cytometer, which can accurately count the numbers of a wide range of cells in a matter of minutes.

It is easy to take this kind of technology for granted today, but the truth is that most of it only emerged in the 1980s.

This explosion of technology and funding directly affected MBC research. For the first time, it became possible to carry out large, controlled studies over time, and receive detailed information about what happens at the level of immune-system cells. It was also possible to determine whether specific psychological states were present before, during or after changes in the immune system, and exactly how such psychological factors translated into chemical and cellular changes in the body.

Finally, it became possible to talk in detail about physical cause and effect. No longer were we forced merely to speculate as to whether there are such mind–body connections; today, we can categorically state that such connections are substantial and measurable and follow logical biochemical pathways.

But there were other developments still to come. Up until late in the last century, we still used X-rays of the brain. It was only around the year 2000 that neuroscientists started to receive all of the work coming out of the use of PET (positron-emission tomography) scans and SPECT (single-photon emission computed tomography) scans, and all the related imaging techniques – particularly those two for my colleagues and me, because they meant that we could now look inside the brain and see what was going on.

As we'll discover in **Part 4**, it was just a few years ago that one of the biggest breakthroughs occurred – one that conclusively proved the existence of MBC. This discovery has also led to the great popularity of meditative techniques and mindfulness, as people have begun to see that reducing their stress levels may have far-reaching consequences. Let's now look at some of those consequences in more detail.

THE BRAIN AND DEEP STATES

When we meditate or mindfully colour in, knit or pray, we gain access to our deeper states. Let's turn our focus to understanding how learning how to achieve these states can improve our brain's performance. We'll start by looking at three critical brain structures relevant to the process of mindfulness (and review others later).

1. The prefrontal cortex

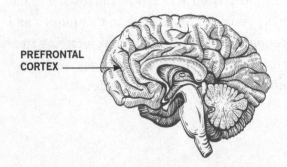

The prefrontal cortex (PFC) can be thought of as the **conductor** of the brain, as it orchestrates thoughts and actions according to internal goals. The PFC is involved in complex planning, personality expression, decision-making, and moderation of social behaviour. When it isn't working properly difficulties can occur. These include attention deficit hyperactivity disorder (ADHD), depression and stress. Studies show that using mindful practices increases PFC activity, and better helps us deal with distracting events and emotions.

2. The amygdala

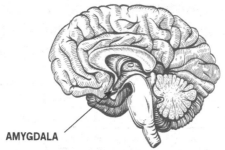

AMYGDALA

The amygdala consists of two almond-shaped structures that, along with the hippocampus, are part of the limbic system. The amygdala is essentially the **security guard** in our brain. It plays a key role in the processing of emotions, and is central to our survival responses – the ones that kick in when we're threatened (or stressed). It tells us when we should be afraid, and it secretes the hormones we need to deal with potential danger.

3. The insula

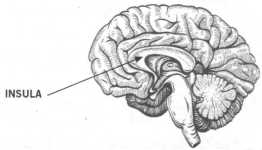

INSULA

This region is often referred to as the **coach**, as its role is to improve core brain and body skills. These include interoception (awareness of our body state), movement, self-recognition, vocalising sounds and music, emotional awareness and perception of time. Studies show that practising mindfulness increases the thickness of the insula, which leads to better awareness of the body, greater skills in dealing with negative emotional experience, and a longer attention span.

How brain cells communicate

The brain has approximately 100 billion neurons, or nerve cells – perhaps more than there are stars in the universe. They communicate with each other using both electricity and chemical substances known as **neurotransmitters**. Across the synaptic gap – the space between neurons – electrochemical 'sparks' fly. On one side of each neural synapse is the presynaptic neuron, which sends the information, and on the other is the postsynaptic neuron, which receives the communication. As neurons 'fire' across the synaptic gap, constant feedback and adjustment by the brain releases further neurotransmitters or inhibits them.

Intricate cell-to-cell communication must occur in the brain for learning to take place. As messages move from neuron to neuron via neurotransmitters, changes can happen within single neurons and among neurons. Changes can also occur to the circuits of interconnected neurons. Learning sensitises a circuit to react in a certain pattern, in order to produce the memory and/or experience again. Over time, the circuit becomes conditioned, so that it only requires a smaller stimulus to set it off.

Neurons are specialised in function and are grouped in the brain accordingly. Two types of neurons that play critical roles in maintaining emotional wellbeing are **mirror neurons** and **spindle neurons**.

Mirror neurons fire not only when we ourselves perform an action, but also when we watch someone else perform the same action. When we see others in the grip of a certain emotion, our brains respond similarly in empathetic resonance. Mirror neurons may actually allow learning through the process of mirroring or imitating another person's emotional and behavioural responses to stimuli. They may also be partly responsible for the transmission of culture, allowing people to absorb the values and emotional expressions of those around them.

Certain social emotions such as shame, embarrassment, disgust and guilt are associated with activity of mirror neurons in the insula. The neural mirroring system could also be an essential mechanism for the sensitive and highly focused empathy between a therapist and subject in hypnosis.

Spindle neurons (also called 'von Economo neurons', after the Austrian scientist who discovered them) are exceptionally large cells that transmit signals from region to region across the brain. They function like air traffic controllers for emotions and seem to be central to social emotions, including our moral sense. They appear to play a key role in our ability to adapt to unstable situations and difficult problems.

Neurological and psychological disorders may reflect problems in either neuron development or the communication between neurons. The abnormal development of spindle neurons can lead to disorders such as psychosis; dysfunction in mirror neurons may be connected with some cases of autism. When the necessary raw ingredients for producing neurotransmitters are missing, or the body's ability to produce neurotransmitters is impaired, this can affect mood, patterns of thinking, and ability to relate to others.

Brain waves

Combining the activity of millions of neurons firing in concert, the brain produces patterns of electrical activity that can be detected on the surface of the scalp. Because of its cyclical, wave-like nature, the electrical activity is commonly referred to as **brain waves**.

When the brain is functioning well, it will use appropriate brain-wave frequencies for particular tasks. **Higher frequencies** suit tasks that require crisp attention; **lower frequencies** are appropriate for activities such as creative problem-solving and sleep. When we meditate, or practise mindfulness by knitting, colouring in or simply digging in the garden, we stimulate these lower frequencies in the brain that cause us to relax and think creatively.

Delta frequencies, 1 to 4 hertz (Hz, or cycles per second), appear during deep, dreamless sleep. During this deep sleep stage, human growth hormone (the hormone responsible for cell and muscle tissue growth) is released, and promotes healing and regeneration. People with ADHD often show high delta frequencies when awake, as do people with brain injuries or various forms of dementia. When people are close to death, they are primarily in a delta brain rhythm, which is a state of suspended feeling and thinking.

Theta frequencies, 4 to 7 Hz, are momentarily experienced when we're waking up or going to sleep. If you suddenly realise you have been walking for some time without even thinking where you are going, you may be in a theta state. (We'll look at this state in more detail later.)

The **alpha** rhythm has frequencies of 8 to 12 Hz. The associated mental state is one of being awake but relatively relaxed. (Again, more on this state later.)

The **beta** rhythm, 12 to 40 Hz, is needed for us to concentrate. A person focused on their work, in conversation, or shopping would be in beta rhythm. People with predominant beta-rhythm activity are action-oriented: movers and shakers. A shortage of beta-frequency activity in the brain has been linked to emotional disorders such as depression, ADHD and insomnia. Beta rhythms are also experienced during periods of high anxiety, stress, paranoia, irritability and mind chatter.

My research has shown that the patterns and repetitions in mindfulness activities like colouring in encourage the brain to produce the more relaxed alpha waves – particularly when these activities are used to help people fall asleep or sleep more deeply.

Gamma frequencies range from 25 to 100 Hz, but are usually over 40 Hz, and indicate intensely focused thought. Research has shown that gamma waves are continuously present during rapid eye movement (REM) sleep. The brains of Buddhist monks who have accumulated more than 35,000 hours of meditation practice have shown these frequencies when meditating on compassion.

There are special relationships between certain neurotransmitters – including the 'stress hormone' noradrenaline (norepinephrine) and the 'relaxation hormones' dopamine and serotonin – and certain brain frequencies. Increases in serotonin lead to increases in the slower frequencies in the theta and delta ranges. Increases in noradrenaline, dopamine and similar neurotransmitters mean that higher frequencies will be stimulated.

The alpha–theta brain

When we meditate, or use mindfulness to colour in, knit or even wash the car, we enter a deep state of brain activity that is little understood. It's similar to hypnosis, and is sometimes called a **liminal space**, or in brain terms, an **alpha–theta state**. It's a period of transition from one state to another, like the space between dawn and twilight.

The alpha–theta state is a transitional state between waking and sleeping, between full conscious awareness and a complete lack of consciousness. It involves the preconscious, the subconscious, and the upper levels of the unconscious. Intuitions, creative ideas, perceptions and impressions pop into consciousness when our mind is in this state. It's a state of deep reflection, in which we lose all awareness of external events.

When achieved, these states have been found to lower stress and increase positive feelings. Modern life lived in a state of stress tends to reduce the number of waking alpha and theta waves that people produce. When people can produce more of these frequencies, they are inclined to feel happier, have better motivation, be more creative, and relate to life events and family in healthier ways. Fear seems to melt away. Deep meditative states lead to less self-criticism, and fewer feelings of shame and negativity.

In the upper levels of alpha–theta, the mind displays **alpha** frequencies; in the lower levels it displays **theta** frequencies. Let's look at both of those states more closely.

The alpha state

Stimulation of alpha waves is caused by repetitive, patterned and controlled tasks such as colouring in.

In this state people can breathe more slowly and deeply, are more flexible emotionally, can think more clearly and intuitively, and can better focus on what they're doing. Increased alpha-wave activity brings greater optimism and motivation and a general state of happiness and wellbeing. These beneficial changes improve the quality of life and also help you live longer. Studies have shown that increasing alpha waves through meditation or mindfulness tends to dilate blood vessels in the brain, increasing circulation, lowering blood pressure and helping to clear arteries.

Athletes who can deliberately keep their mind calm and focused tend to perform at or near their personal best. Developing the ability to go into 'the zone' (the alpha state) at will is also a strategy for ageing well. Younger people tend to produce alpha waves easily, while older people often show a decrease in these waves. However, older people can be trained to increase their production of alpha waves using mindfulness, which helps them relax and can actually rejuvenate their brains. In one study, seventy-year-old people who were taught to achieve an alpha state had renewed energy and motivation, and showed brain-wave patterns commonly associated with thirty-five-year-olds.

Because the sensation of hunger activates the more focused beta state in the brain, controlling the appetite with alpha waves is one key to weight management. On the downside, people with overly high alpha states often end up anxious when trying to focus.

The theta state

Theta brain waves produce a state of deep relaxation. People in theta state often experience hypnagogic images (as they fall asleep) and hypnopompic images (as they wake up) from the unconscious mind. There are many examples in which people dream about solutions to problems or have 'waking dreams' filled with hypnagogic imagery. As they begin to awaken, sudden flashes of insight and creativity occur. This brain rhythm is also present in deep meditation and in deep hypnosis.

Through the theta state, we can reach the higher self or 'true self'. The return to a more resilient and vulnerable state, similar to returning to a younger age, can open the door to dramatic change. Gaining access to deeper levels of awareness and inner resources may allow a person to develop a 'psychological immune system' so that they can begin to release the mind's baggage. People with dominant theta rhythms are likely to be highly intuitive.

The characteristics of the theta state include:

- A feeling that time has slowed down or disappeared
- A slowing of mental activity and the disappearance of mind chatter
- The feeling that you will recover from an illness
- The disappearance of body awareness and sensations of pain
- A transcending of core beliefs about the limits of what is possible in life
- The discovery of unique solutions that would not have been found in other states of consciousness
- The extinguishing of fears
- Decision-making and the resolution of difficult issues guided by suddenly perceived insights
- The transformation of problems that seemed insurmountable into interesting adventures
- The stimulation of creative ideas or information (creative people such as artists, musicians and writers have used theta state to engage their creativity)
- Healing of unresolved emotional issues
- Awareness of stress and the ability to reduce it, as well as the reduction of levels of medications used
- Better tolerance of life obstacles
- Greater compassion, awareness, detachment and inner security

Studies of deep states

Though it began earlier, scientific study of meditation became widespread in the 1970s and 1980s. Several investigators analysed the EEG recordings of Zen monks who practised meditation regularly. As the monks went into meditation, they passed through four stages. The first stage showed the appearance of alpha waves; the second stage revealed that there was an increase in the amplitude (strength) of the alpha waves. The third stage showed a decrease in alpha, followed in the fourth stage by an increase in long periods of theta. The longer the monks spent in each session of meditation, the more the theta frequency was produced, although the monks' minds were still completely alert. Early deep-state researchers concluded that it might be possible for anyone to learn how to control the mind and change brain states at will.

Dynamic meditation is a form of mind training that emerged in the 1970s and is still popular today. It involves a descent into the deeper realms of consciousness, where information can be gleaned about issues ranging from health to financial success. At the core of dynamic meditation is the suggestion that we come into the world through the delta level and leave at the same level. It is believed that with mental rehearsal in an alpha state, we can achieve our goals and reprogram our brain for success.

Stages of deep state

Neuroscience has expanded the monks' four stages to six discernible stages, ranging from normal consciousness (beta state) to profound (alpha–theta) states and back again.

The first stage is called **settling**. Here a person enters an initial state of relaxation (alpha state). Some people become frightened when they first relax, because it feels so foreign. Occasionally at this point people will report a feeling of falling.

The second stage is **deepening**. The person's muscle tension begins to decrease; breathing becomes easier and deeper. Often the person will sigh deeply; when this happens, the journey into theta has begun.

Following this stage is a **dissociation** of the conscious mind from the unconscious mind. The unconscious mind quietens, and the person often begins to experience trance phenomena: a loss of bodily sensation or inability to feel pain or anaesthesia; heaviness or warmth; immobility and rigidity; and a feeling that time has slowed down or stopped. Usually the person is completely quiet at this point. Their muscles become flaccid due to relaxation, and often REM begins.

As the state deepens and the person heads towards sleep, they reach the **theta crossover stage**, where the ratio of theta to alpha waves shifts in favour of theta waves. We can discern when a person is in this stage by the quiet and still state of their body. At this point the person perceives images transcending time, culture and language. Sometimes, voices of loved ones are heard, or an awareness of their presence is felt. In this state of reverie, the person lightly notices the images they see and does not try to hold on to them. Being in a receptive state allows the images to flow, whereas if the person attempts to hold on to them they will disappear. Later, in processing the images, the person can ask the unconscious for an associated meaning as they ponder the message.

The next, and deepest, stage is **deep theta**, in which the mind feels quiet and empty. To enter this state, the person must feel safe. It is in this state that movement into healing and regeneration takes place. In this state, the individual moves beyond conditioned beliefs and can perceive creative solutions that will allow them to achieve what they previously thought was impossible.

The sixth and final stage is **reorientation** to the person's immediate surroundings. In this stage, the person gradually begins to reconnect with their body and shifts to a more active state. This reorientation needs to be done slowly to avoid potential headaches or discomfort. (It's possible for a person to encounter frightening images and upsetting emotions in theta state. This may occur if they remain too long in a state with a dominant brainwave of less than 3 Hz, without proper preparation or ability to lift out.)

Implications of deep states

It's healing to separate from our normal surroundings and shift into a state of timelessness and silence for a period of time through mindfulness or meditation. Deep reflection leads to the emergence of a sense of meaning and life purpose.

People who are able to **monitor and control their own behaviour, emotions and thoughts** tend to be the happiest. Those who can learn how to notice reactions without acting on them have the best management over themselves. Those who are carrying heavy armour after being hurt in the past often have difficulty with this. They've been overly criticised, controlled and forced to act in ways that don't suit them. Time spent in deeper states can unhinge this armour, break through old patterns, and open them up to a better quality of life.

Some people are so out of touch with their own bodies that it's difficult for them **to be aware of subtle physical differences**. Deep states can help with this. I often encourage people to notice differences between feeling depressed and feeling fatigued, or feeling anxious and feeling hungry (as a result of low blood sugar). When a person feels bad, the categories of 'bad-ness' become collapsed into one, and everything is negative. When a person is depressed, the world is perceived through a grey lens, and it's difficult to notice small breakthroughs of colour. Deep states will improve their outlook on life.

Deep states also tend to alter a person's **understanding of past events**. People are often able to forgive those they perceived had harmed them. They may 'hear' and 'see' things in deep states that can help them solve a past dilemma. Or they may ask the unconscious mind a question about some personal issue; often the answer will give them a surprisingly different perspective. By healing issues from the past, they can focus on the present and the future, free from mental clutter.

Working with deep states through techniques such as mindfulness and colouring in can be particularly helpful in **dealing with past hurts**. When a person discusses painful issues in their past, they are often flooded with anxiety, and their thinking is affected, leaving them in an even worse situation. Time spent in deep states effectively 'cools' the brain and may allow the person to view what happened in the past from an empathetic adult perspective. The calm state will continue after the person returns to normal consciousness. This can help them gently discharge the negative emotions from their memory.

Newer techniques involving **floating in sensory deprivation tanks** have been found to help people achieve even deeper states of alpha and theta. Several sessions in one of these tanks have cleared many maladies. Overweight people have lost weight, rheumatoid arthritis pain has decreased significantly, depression has been lifted, anxiety has been calmed, immune responses have increased, and people have seemed happier over all.

As research into the implications of deep states continues, who knows what other exciting findings will emerge?

CONSCIOUSNESS AND MEMORY

Conscious – subconscious – unconscious

Most of what is contained in an individual's mind is not easily accessible. Most of us are only aware of a very small part of it.

For example, when you forget where you put your keys, the memory of where you put your keys down is simply not available. However, after a period of thinking, you may remember where you put them. In the same way, much of what you've learned and experienced in your life is 'forgotten', i.e. it becomes unconscious. Not remembering doesn't mean that it's not stored somewhere in your mind.

Sigmund Freud was the first person to name the conscious, subconscious and unconscious levels of the mind, and most psychology is based upon his work, directly or indirectly. Yet many aspects of consciousness and the mind are still a mystery.

Your **conscious mind** contains:

- All the thoughts and memories you are currently aware of and can quite easily remember

- Your logic and reason

Your **subconscious mind** contains:

- Things that are difficult to remember, but which you can remember with a little effort and concentration

Your **unconscious mind** contains:

- Past memories that you can't remember
- Attitudes and beliefs you have that you are unaware of

The conscious mind holds current information and currently perceived emotions, moods and attitudes. It can hold about four pieces of information at one time if they are not complex, otherwise it can hold only one piece at a time. However, the conscious mind can also retrieve stored data almost instantaneously. It constantly uses past experiences to direct present emotional states and behaviours, and to evaluate potential future choices.

There is sometimes confusion about the meanings of the words 'subconscious' and 'unconscious', as they are often used interchangeably. Strictly speaking, the correct word for all memories and beliefs that you are not aware of, and which are stored deeply in your mind, is the unconscious.

The term **subconscious** more accurately describes thoughts and feelings that are *temporarily* just under the surface of conscious awareness, and can be remembered fairly easily. The earlier example of forgetting your keys, then remembering where you put them a few minutes later, is an illustration of the *subconscious* level of awareness – a temporary forgetting that is easily remembered. The subconscious may be viewed as the thin top layer of the unconscious, or the layer between the conscious and unconscious layers of your mind.

A good example of the difference between conscious and unconscious parts of your mind involves the formation of habits and skills, such as learning to drive a car. How many times have you driven from one place to another and realised that you couldn't remember the journey because you were daydreaming? Yet there was a part of your mind that knew exactly what to do – when to change the gears, apply brakes, stop, start, speed up, slow down, and a whole range of other activities – all of which you had no conscious memory of doing.

Similarly, the memories of the important and unimportant events of our lives slide into the unconscious part of our mind, and continue to influence us all the time, even when we are unaware of this influence.

The challenge in working with the mind lies in past decisions and events that influence present-day feelings, thoughts, attitudes, beliefs, values and actions, even though you're not necessarily aware of them – just as you automatically *know* how to drive even when you're thinking about something else. (We'll look at this more closely in the next section, **Memories and brain change**.)

As you begin to ask questions like 'What must I believe about myself to behave this way?', you begin the process of bringing these unconscious beliefs into your consciousness. Initially you may just have a feeling about them, or dream about them. When they're 'just on the tip of my tongue', they've reached the subconscious level. Soon you'll become consciously aware of these beliefs.

This is why it takes time to work through life's issues. You have to first assume that there may be something in your mind that is causing some behaviour. You may have no idea of what it is – it is unconscious. By persistent focusing and effort, the thought or belief surfaces, first into the subconscious level of your mind, and then into conscious awareness.

Sometimes the presence of unconscious beliefs is obvious, such as when you sincerely want to stop smoking, start a diet or exercise, but you cannot seem to get it right, no matter how hard you try. In these cases, you are aware of the *effects* of the unconscious beliefs, but don't know what these beliefs are or how to deal with them.

Most of the time these unconscious beliefs are not at all obvious, or they present themselves as acceptable 'reasons' why something is not possible. A typical example is when someone is 'too busy' to take care of their health, or 'forgets' their appointment with a doctor or therapist. Many people are truly unaware of how their unconscious beliefs influence their actions.

'But,' you say, 'how do I know that my so-called unconscious mind exists in the first place? It would be just as easy to say that there's no such thing as the unconscious, and that my behaviour and feelings are just "natural"!'

Excellent question! The example of forgetting your keys, someone's name or some object, then remembering it soon afterwards, seems to indicate that your memory of something does not just dissolve when you forget. It seems to indicate that it temporarily 'goes somewhere' – almost like a filing system – until you 'find it' again.

You may also have noticed that certain smells, images and sounds can spontaneously bring back a memory of the past – even from your childhood. Where were these memories before you remembered them again? The only logical answer I can offer is that they were stored away in the unconscious mind.

'OK, so memories are stored away somewhere,' you say, 'but are *all* memories stored away? Surely some things are permanently forgotten?'

Apparently not. There is some fascinating research that indicates even conversations that occur while you are in a coma are stored away in the unconscious mind, and that these memories can be recalled with techniques like hypnosis. The unconscious mind holds memories of the entirety of a person's experiences from the beginning of life. It also processes a great deal of input outside conscious awareness. It has an 'internal search' function that enables it to come up with the best solution at any given time, and it continues searching for the best answer into the future. This is often called the **adaptive unconscious**.

It can size up people's emotions, character and intent quickly and accurately, and allow us to make snap intuitive decisions. It may give us a sudden flash of insight or a spontaneous 'knowledge' that something is about to occur. (Think of a policeman who suddenly shouts to his colleagues to take cover as a suspect approaches, before any weapons are seen.)

These unconscious processes form a huge body of resources for the conscious mind to tap into for problem-solving and other complex tasks.

Much of the unconscious mind's activity is based on an ability to recognise familiar patterns – the same patterns we use when we practise mindfulness. My research has found that when people are asked to think about a complex decision, they often make poor choices. But if they are asked to do some colouring in before making the decision, they make better choices and are more satisfied with the end results. As we saw in the previous chapter, a relaxed state of mind achieved through mindfulness allows the unconscious to reorganise ideas, issues and perspectives, which can lead to better decision-making.

We often experience a thought followed by an action and assume the thought *caused* the action. However, it may be that the thought and action come from another unconscious process that precedes both. If so much of what we think and do happens at a unconscious level, effectively we're already doing these things before we thought of them. When neuroscientists put MRI (magnetic resonance imaging) machines on two people who are looking at each other and thinking and talking, they can see the frontal lobe activate prior to any conscious stimulus from the other person; in other words, there's an answer in the brain *before* a person has consciously thought about it. Which begs the question: 'Do we actually have any control over anything we do?' The science is saying that maybe we don't.

Memories and brain change

An **implicit memory** is a memory that you don't realise you're receiving from the past. These memories are tucked away in our unconscious but, as we've just seen, have a significant influence on our behaviour in the present. Many are useful and necessary, as they are recollections of things you know how to do that you don't need to recall consciously.

For example, when you first learned to ride a bicycle, you were very aware of putting your feet in the right place and keeping your balance while simultaneously pedalling and steering properly. This felt very complicated for a while, and it probably took quite a few attempts, and several falls and scraped knees, before you got all the various actions coordinated. However, if I were to ask you how it is you know how to ride a bicycle, you probably wouldn't remember how difficult it was to learn the skill; instead you'd reply, 'Well, you just kind of sit there, balance and hold the handlebars – it's easy!' Implicit memory helps you remember how to ride a bike without consciously feeling like you are having a memory of learning to ride. (This is similar to the example we looked at earlier of being able to drive a car without consciously remembering what's involved.)

But implicit memory can be deeply troublesome if you experienced intense emotions in the past. The implicit memory of these things can emotionally hijack you in the present even if you're not aware it's happening.

Implicit memory, however, is not the only player in town. We also have **explicit memories**, the memories we *know* we're having. They're encoded in the brain's hippocampus and combine with our implicit memories to ensure we're aware of what is happening to and around us.

Now, here's the really interesting part: the hippocampus often shuts down when intense emotions are experienced and prevents explicit memories from being stored. If the hippocampus stays online during an emotional experience, the memory seems easier to remember.

Fear often gets encoded as an implicit memory. Making these types of implicit memories more conscious can help to integrate them with explicit memories, which can decrease their power over life in the present.

The more energised we feel, the more likely we are to rely on our explicit memories. The brain needs to replenish its energy through sleep, but mindfulness can also be used to energise us and heighten our ability to use explicit memory.

Tiredness and lower energy levels mean we're more likely to allow implicit memories to invade our life. We'll be more likely to do things 'the way we've always done them', and the brain won't look for new pathways.

Neuroplasticity is the ability of the brain to change itself by learning and creating *new* pathways. To imagine how these neuronal pathways are 'worn' into the brain, look at the image on the next page.

When we learn something new, the brain 'cuts' a neuronal pathway between the pieces of information it receives (like a new path to your house). To maintain the path we need to keep repeating the task (keep walking on or mowing the same path). Such pathways are at work when we automatically respond to things without stopping to think.

If we do something new or in a different way, we start to change the pathways; this is neuroplasticity.

Just like our brain needs energy to use explicit memory, it requires energy to change. MRI studies, SPECT scan studies and EEG studies confirm that mindfulness is an excellent way to trigger positive neuroplastic changes. These studies show improvements in ability to control thoughts and emotions, mood, wellbeing, self-esteem, concentration, sleep, health, memory and much more.

Does fear or anxiety play a role in YOUR life? Are past memories holding you back? Mindfulness can change not just the way your brain works, but its actual structure as well.

BALANCE AND CONNECTION IN THE BRAIN

Homeostasis (balance)

To enable us to function at our best we need the energy flows between our brain and body to work efficiently. When your brain is performing at its best, it means there is homeostasis (balance) between the four key brain regions. If there is a lack of balance and one of the four areas becomes dominant, you may experience the following signs:

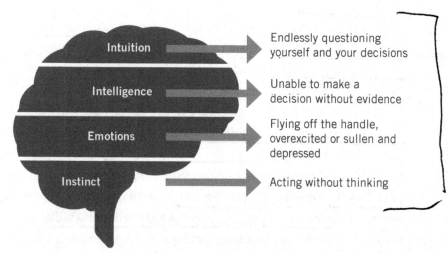

When the brain becomes more relaxed through mindfulness, your system corrects itself and returns to balance.

Remember, mindfulness is not a point you get to, it's a journey: you're trying to get from 60,000 thoughts a day to just being in the moment and not having *any* thoughts.

Coherence (connection)

Another important concept is **coherence**, a measure of how well various areas of the brain connect with each other. Low coherence means two areas will be locked together while functioning. For instance, when the coherence between Broca's area, which is responsible for speech, and Wernicke's area, responsible for interpreting language, is too high, speech disorders can result.

As we saw in the previous chapter, the neurons in our brain generate an electrical impulse when they fire. Collectively, they generate an electrical field. At the same time all the other cells in the body generate minute but measurable electrical fields too. Because every movement in the body produces these microcurrents, all cells are connected in terms of overlapping electrical fields, which also extend *beyond* the body.

Organ cells tend to fire in concert, so organs generate much stronger fields. For example, the heart generates small electrical waves (measured in millivolts) that can be detected by ECG (electrocardiogram).

It is well known that a healthy heart speeds up with every breath in and slows down with every breath out. Each time you exhale, your brain sends a signal down the vagus nerve to slow the cardiac muscle. With each inhale, the signal gets weaker and your heart revs up. Plotted over time, these variations generate a pattern called **heart rate variability** (HRV). This variation is driven by the interplay between the sympathetic and parasympathetic nervous systems: the sympathetic nervous system (the danger or stress system) speeds up the heart rate, while the parasympathetic system (our everyday working system) slows it down.

Although our understanding of HRV is far from complete, it seems to be sensitive to acute stress. In the laboratory, high mental functioning and complex decision-making have been shown to lower HRV. HRV has also been shown to decline as we get older, and a decreased HRV is often an early sign of illness. By keeping HRV within optimal levels, a person can lower their stress and increase their overall health.

HRV can be increased by using an exercise strategy of high exertion and recovery – for example, several brief bursts of high-speed running alternated with a few minutes of rest. During the rest phases a mindfulness technique such as colouring in could be used to heighten the impact.

HRV is also influenced by emotional states. Positive feelings such as compassion are associated with coherent patterns in the heart's rhythms. When we feel negative feelings such as anger, our heart rhythms degenerate into less ordered patterns, and our body feels stressed. Studies done on the risk of developing heart disease have shown that *both* people who vented their anger *and* those who repressed it tended to significantly increase their risk.

My own research has found that when people intentionally shift their heart rhythm to a more coherent rate, their emotional state improves. Coherence appears to be a more powerful physiological state than relaxation. Using colouring in or other mindfulness techniques, people can learn to slow their heart rate and maintain alpha brain waves, giving them a sense of calm. Improving heart-rate coherence also involves learning to breathe properly – which we'll discuss in **Part 3**.

The rhythms of our universe

Because the body's electrical field extends beyond the skin, other people may perceive the electromagnetic fields that are generated by someone's coherent or chaotic heart patterns. In essence, a person's emotional state can affect others.

Although a few people say they can perceive these fields consciously, we often register another person's electrical field at a subconscious level. It's what we might call someone's 'vibe': 'I get a good vibe from this person.' Research has shown that when a group of people start singing together at a concert, their heart rhythms all come into sync. And when two people live together or are connected by strong emotional ties, the electrical patterns of their brains and hearts may become naturally aligned. In a similar vein, it's long been known that the monthly periods of women who live together often become synchronised.

Even though it's mainly unconscious, this ability of people to come into a biomagnetic alignment has ground-breaking implications. In fact, scientists have proposed that not just the heart, or even the body as a whole, but everything in the universe moves in waves.

Rhythmic patterns can be observed at all levels: from atomic patterns, to the vibrating structures of molecules, to the wave patterns found in complex organisms. People are awash in the rhythms of their own pulsating cells, fluctuating hormones, and cycles of growth and maturity. They are also enveloped in our world's rhythms, cycles and waves: **ultradian rhythms** of 90- to 120-minute cycles (during which our bodies slowly move from a high- to a low-energy state), **circadian (solar) rhythms** of twenty-four hours, weekly cycles, monthly (lunar) cycles, and rhythms produced by sound and light.

Circadian rhythms, for instance, are related to hormonal changes and weight gain. Research suggests that a treatment for obesity might include changing the circadian pattern of light and dark. Exposure to fourteen-hour periods of darkness can trigger hormone release and encourage deeper and more restful sleep.

Ultradian rhythms seem to relate to the periodic release of certain hormones that affect attention span and hunger. Experiments have shown that people head for the refrigerator or the coffee pot roughly every ninety minutes partly in response to the natural ultradian cycle.

It seems reasonable to suggest that we become stressed or ill when we stifle these waves within our body. In other words, maintaining the waves inside and outside the body is a key to maintaining health. For instance, when our heart rhythms come into sync, it's incredibly powerful in a health sense. When we're lonely we have very poor health; when we have family and friends and social networks, we have very good health. This is all linked to our natural heart connections.

Mindfulness techniques are a key to maintaining these waves within and outside us. In **Part 3**, we'll look at practical exercises we can use or adapt to achieve mindfulness – whatever our tastes and lifestyle.

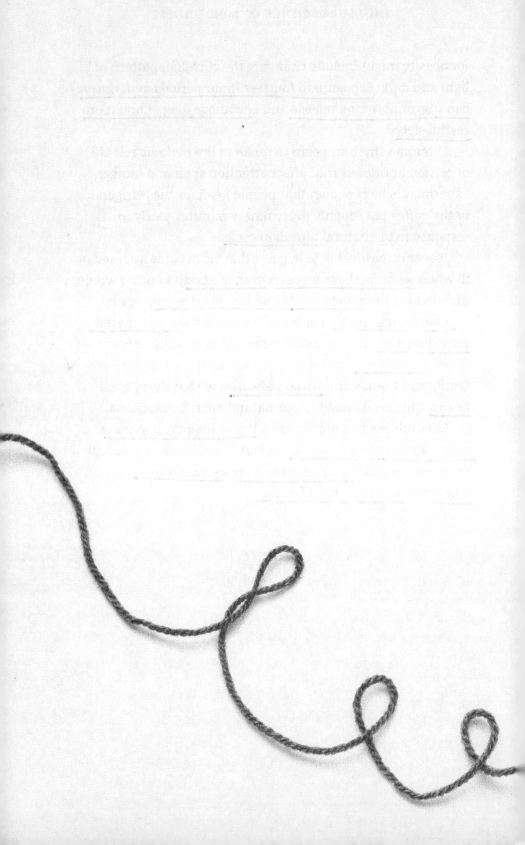

PART 3

MINDFULNESS IN EVERYDAY LIFE

CORE SKILLS

In mindfulness we need to focus on **three core skills** to help us work more effectively while colouring in or doing another activity. These are:

1. Setting our intention

2. Cultivating awareness

3. Regulating attention

Let's look at each of these and how they affect our brain's ability to focus and concentrate.

1. Setting our intention

Setting intention is a basic step in any activity. Your intention is **what you wish to achieve from an action**. In mindfulness, intention refers to what you are choosing to pay attention to. For example, your intention might be to go offline for five to ten minutes and pay attention to your activity while slowing down your breathing. Being mindful involves bringing your attention back to your intention, over and over again.

EXERCISE
Setting our intention

- Decide your intention – for example, to spend five to ten minutes relaxing while colouring. (You can easily adapt this to another mindfulness activity.)

- Identify this intention by taking out a colouring book (such as a *Colourtation* book) and beginning to colour. (To enhance concentration, you could listen to music through headphones – or try the 'alpha sounds' at soundcloud.com/colourtation, which go for exactly six and a half minutes.)

- Keep awareness of your intention present in your mind.

- Remind yourself as soon as you become aware of a thought, feeling or distraction that your intention is to focus on your colouring. Then shift your attention back to your task.

- Check in periodically to ensure your thoughts and actions remain consistent with your intention.

'Being mindful involves bringing your attention back to your intention, over and over again.'

2. Cultivating awareness

Being aware of awareness is a revolutionary idea for many people. It's about noticing what's arising as it arises. This includes awareness of thoughts, feelings, body sensations and physical surroundings. It involves paying attention to what is happening in this moment, and acknowledging and dismissing distractions.

The goal is to remain aware without trying to change anything. In this case, just be aware of your colouring. Observe and simply accept what you observe.

Awareness is the first step in being able to change unwanted patterns of thought and behaviour.

EXERCISE
Cultivating awareness

- Pause for a moment.
- Notice what's arising as it's arising. Pay attention to thoughts, feelings, body sensations, surroundings. Just be aware without trying to change anything.
- As distractions occur, remember your intention and bring your attention back to what's arising.

3. Regulating attention

Regulating attention means **paying attention on purpose**.

We live in a very distracting and distracted culture. We often feel overwhelmed and overloaded, with many things competing for our attention. Practising mindfulness increases the brain's ability to regulate itself. Being able to regulate our attention will improve our concentration, memory and overall mental clarity. Neuronal pathways are created and strengthened by the repeated practice of calming the mind and paying attention to something on purpose.

EXERCISE
Regulating attention

- Set your intention to pay attention. Select colouring as the thing you will be paying attention to.

- Now just notice everything on the page you have selected to colour. Notice if you are being distracted by a thought, feeling, body sensation or something in your environment. Acknowledge the distraction and dismiss it without judging.

- Return your focus to your colouring. Notice the details of the image you have chosen to focus on. Notice how it looks and what feelings it evokes.

- Continue this process for five to ten minutes then stop.

MINDFULNESS ACTIVITIES

Now let's look in more detail at the range of activities you might practise to develop mindfulness, and how you might refine your techniques.

Everyday activities

As we've seen, you can apply mindfulness techniques to a huge range of other tasks – even an everyday activity like **brushing or combing your hair**. The task you choose just needs to feature the three key ingredients we heard about earlier:

1. Pattern

2. Repetition

3. Control

The key here is to turn off the busy, distracting chatter going on in our brains so that we can focus on being in the present.

EXERCISE
Practising mindfulness while brushing or combing your hair

- Remember throughout this activity to notice when your attention has wandered and remind yourself of the activity you should be paying attention to: brushing or combing your hair.

- Stand in front of a mirror. Look at yourself in the mirror, take a deep breath and say 'Relax' to yourself as you exhale.

- Pick up your brush or comb. As you grasp the handle, start to pay attention to how it feels in your hand. Is it warm, cold, soft, hard, smooth, textured, slippery or sticky?

- Clean the brush or comb by rinsing it under the tap. As you turn on the tap, notice how the tap feels on your hand and fingers. Is it warm, cold, soft, hard, smooth, rough, slippery, sticky? Is it shiny or dull? Is it dry, or covered with water?

- As the water starts to run into the basin, look at it for a moment. What does it look like? Is it a steady stream? Is it dripping or rushing out? Is it gurgling straight down the sink or starting to fill the basin?

- Place your brush or comb under the water and notice how your hand feels as the water flows over the brush or comb. Is your hand wet? What sound is the running water making?

- Pick up your brush or comb container (if you have one). Notice how much it weighs. Pay attention to how it feels in your hand. Is it warm, cold, hard, soft, smooth, rough, slippery, sticky?

- Run your fingers over your brush or comb. What kind of fragrance does it have? Is it a fresh smell? Are the bristles

continued

> - of your brush or the teeth of your comb hard and stiff, or soft and flexible?
> - Notice how it feels as you brush or comb your hair. Is there a tingling sensation as you move the brush or comb across your head? Is the movement easy or difficult? Is your hair moving into the shape you want?
> - Run your fingers through your hair and notice how your scalp feels. Do your fingers feel different from the brush or comb?
> - Pay attention again to your brush or comb, then put it away after you have finished.
> - Look at yourself again in the mirror. Take a deep breath, relax and enjoy the feeling that comes from a brain that has been absorbed in mindful activity.

What skilled, semi-skilled or unskilled activity can YOU think of that will help you apply the principles of mindfulness to YOUR life?

Body awareness

As mentioned in **Part 1**, a popular mindfulness exercise is carrying out a 'body scan'. This involves focusing your attention on each of your body parts in turn, noticing when you feel something, and sending positive and calm thoughts to each area of your body.

Practising this regularly helps you to accept and work more effectively with your body, respond to its sensations and develop deeper states of relaxation. In turn, this can help prepare you for other mindfulness exercises, including meditation.

EXERCISE
Body scan

- Lie down or sit in a comfortable position, making sure you are warm and can relax.

- Start breathing in slowly through your nose. Count to three, then breathe out through your mouth even more slowly while counting to five. Return to normal breathing.

- Focus your attention entirely on your left foot. Think about the toes, the heel, then the bottom of your left foot and the top of your left foot. Pay attention to how each area feels. Then move up to your left ankle. Notice how it feels. Is there any pain there? Is it cold or hot? Does it feel light or heavy? Accept the sensations as you become aware of them.

- Now switch your attention to your right foot: the toes and heel, the bottom of your right foot, the top of your right foot. How does each area feel? Then move up to your right ankle. Is there any pain there? Is it cold or hot? Does it feel light or heavy?

- Shift your attention to your right leg, starting at the bottom then moving up to your knee, thigh and all the way to your hip at the top of your leg. Notice if your hip feels tight or relaxed, warm or cold, light or heavy. Send positive and warm thoughts to your whole right leg. Repeat for the left leg.

- Next, pay attention to both your legs, from your toes up to your hips. Be still, breathe and send your legs some kind thoughts. Breathe gently into your legs.

- Move your attention to your stomach region. Just observe what is there. Notice how it feels. Let it be the way it is. Send kind thoughts to this region.

continued

- Focus now on your back, starting at your lower back and moving all the way up to your shoulders. Notice any sensations present in your back. Send warm thoughts of relaxation to your back. Stay still for a moment, noticing everything about your back.

- Switch your attention to your fingers, thumbs and wrists. Think about what your hands are carrying. Send kind thoughts to your hard-working hands. Then focus on your arms, from your wrists all the way up to your shoulders. Just notice what is there.

- Pause to focus on your breathing. Pay attention to your neck and throat. Swallow and notice how your neck and throat feel. As you do this, send thoughts of health and healing to your neck and throat.

- Now pay attention to your face: your chin, your mouth, your cheeks, your eyes, your eyebrows, your forehead, your ears. Take a moment to sense what's there. Notice everything without trying to change it. Send positive thoughts. Let yourself smile.

- Bring your attention next to your head, including your hair and scalp and your brain inside your skull. Monitor the activity inside your head and mind. Send positive thoughts. Connect to your inner self and all its knowledge.

- Finally, take a deep breath right down to your belly, and fill your whole body with calm, relaxing energy. As you blow the air out, gently let go of anything that needs to go.

- Open your eyes slowly and bring your attention back to where you are.

Colouring in

As discussed earlier, my studies have shown that colouring in is one of the most effective and easily practised mindfulness activities. For maximum benefit, try to do this activity at least twice a day, away from your computer or other distractions. If you want to, put in your earphones and start playing your favourite music or the alpha sounds at soundcloud.com/colourtation.

EXERCISE
Mindful colouring

Before you start colouring, try to picture a colour that will help you stay focused. Use a colour that is significant or pleasing to you. Experiment with different colours until you find the best one(s) for you. Let your colour choices just come to you. The list below shows some of the emotions I've found are commonly linked to certain colours by patients using my *Colourtation* colouring books.

White	Frustrated, confused, bored
Black	Harassed, overworked, tense
Grey	Nervous, strained
Dark blue	Happy, passionate, romantic
Blue	Relaxed, calm, at ease, loving
Pink	Fearful, uncertain, questioning
Purple	Sensual, clear, purposeful
Red	Excited, energised, adventurous, ready to go
Orange	Stimulating ideas, wanting, daring
Brown	Jittery anticipation, restless thoughts
Dark green	Inner emotion charged, somewhat relaxed
Mid-green	Average, not under stress, reasonably active

continued

Light green	Unsettled, cool, mixed emotions
Yellow	Imaginative, feeling OK, wandering thoughts

- Choose any picture you like and just focus for five to ten minutes on colouring. You do *not* need to finish the whole picture – you can come back to it next time. All you need to do is let yourself become immersed in the activity and enjoy it for what it is – colouring! Concentrate on relaxing and watching your creativity come through. The brain loves this.

- Remember, your **intention** is to relax for five to ten minutes. Be **aware** of what you are doing and focus your **attention** only on the image you are creating.

- Take as little or as much time as you can manage. If you colour in at least once or twice a day, or whenever you feel stressed, you'll soon train your brain to relax.

Crafts and hobbies

Many of the things we now regard as leisure activities were once necessary to our survival. We needed to garden to grow our own food; we needed to sew to make our own clothes. More recently, though, many of these pursuits have made a resurgence, as people discover the relaxing properties of simple, repetitive tasks, requiring varying levels of skill.

Knitting is one of the crafts that's been making a comeback – partly because people are recognising its therapeutic mindfulness benefits.

EXERCISE
Practising mindfulness while knitting

- *Remember throughout this activity to notice when your attention has wandered and remind yourself of the activity you should be paying attention to: knitting.*

- *Find your knitting and sit in a comfortable chair.*

- *Pick your wool up in both hands. Pay attention to how it feels. Is it thick, thin, soft, scratchy, smooth or textured?*

- *Smell the wool. Take the aroma in. Is it fresh, musty, lightly scented or highly scented?*

- *Put the wool in your lap and pick up your needles. Pay attention to how they feel between your fingers. Are they smooth, textured, warm, cold, thick, thin, dry or greasy? Do they look shiny or dull?*

- *Before you start to knit, look at your knitting for a moment. Is it just started, almost finished, neat, untidy, loose or tight? What kinds of stitches have you used?*

continued

- Start to knit. Notice how your hands feel as they begin to move with the needles.

- See how the wool moves effortlessly onto your needles and begins to create a pattern. See the wool's pattern grow as you continue to move the needles.

- Sense your breathing. Feel the relaxing breaths out as your wool effortlessly forms your creation.

- Your fingers are busy. Watch them work together to move the needles. Focusing on your fingers, consider: are they tense, relaxed, rough, smooth, old, young, stiff or flexible?

- Watch the wool ball shrink as you effortlessly continue to move your needles in unison.

- Notice how your creation takes shape and how good you feel.

- As you finish your knitting and put down your wool and needles, give yourself a GREAT BIG SMILE.

> 'People are beginning to recognise the therapeutic benefits of traditional crafts.'

Mindfulness outdoors

You can continue to practise mindfulness even when you are out and about. Let me show you how with the following three exercises.

EXERCISE
Being outside

- The first thing to do if we wish to practise mindfulness outside is to take in our surroundings.

- Look at the sky. Take in the colour. Is it cloudy or clear? What do the clouds look like? Is the sun bright and shining? Or are the clouds covering it? Is it warm or chilly? Windy or calm?

- Look around you. What can you see? Are there shrubs and trees? Look closely at the trees. Are they covered with leaves or bare, or somewhere in between? What colours are the leaves? What colour are the branches? Are there flowers or buds? Is the tree still or moving? What are the shapes of the leaves? Are they big or small? Touch them if you can. Are they scratchy or smooth?

- Now slowly inhale and notice what you can smell. Is there a fragrance or odour? Is it sweet or sour? Pleasant or unpleasant? Is it natural or something else? Does it remind you of something or perhaps take you back to another time and place in your life?

- Can you see any grass? Look closely at it and try to take in its colour. Is it dried out and brown, or lush and green? Is it neglected or carefully looked after? If you're sitting on the grass, look down and touch it. Feel it between your fingers, rub some leaves together. Is it moist or dry?

continued

- Are there any flowers nearby? If so, what are their colours and shapes? If you can touch them, how do they feel?

- What else is around you? Are there any rocks in view? If so, take in their shape and colour. Again, touch them if you can. Feel their surface. Are they clean or dirty? Sharp or smooth?

- Is there any water near you? A pond, river, ocean or just a puddle of water left after a shower? Pay attention to the water. Is it calm and still or moving? What is the colour of the water? If you are at the beach, can you smell the salt from the water?

- Let's listen. What do you hear? Just the wind? Birds? People? Or perhaps the sounds of the city: cars, trucks, buses, trains, horns, sirens, planes?

- Pay attention to the temperature. Is it hot or cold? Warm or chilly? Is it breezy or still?

- Finally, ever so carefully, start to bring your attention back to your daily activities.

'Practising mindfulness outdoors can help you feel connected to the earth and be fully present in your body.'

EXERCISE
Mindful walking

- While we walk we can practise mindfulness. We just need to focus on the movement of each step and on our body as we move.

- Since walking is for most people a daily activity, mindful walking is a great way to increase your quota of mindful experiences. You can do it just about anywhere and anytime; it can help you feel connected to the earth and be fully present in your body; and, perhaps most significantly, it can calm the busy chatter of your mind and provide clarity.

- To try it out, find a place where you have room to move back and forth, at least five to ten steps in length. You can practise inside or when you go for a walk outside, or even on a treadmill. Try to keep your hands still, either behind your back, or at your sides.

- Before you start walking, focus your attention on your breathing. Notice how your breath flows effortlessly in and out.

- Stand and balance your weight evenly between both feet. Now feel the sensations of standing. Bring your attention to your body. Be conscious of contact with the ground, of pressure and tension. Pay attention to how it feels to stand. You should begin to notice how all the parts of your body participate in standing, including your feet, ankles, hips, stomach, chest, back, shoulders and your arms.

- Time to lean gently to your right side. Can you notice what feels different as you put all your weight on your right side?

continued

Pay attention to the feeling in your right foot, leg and hip, and your back, neck and arms. Pay attention to the feeling in your left foot, leg and hip. Notice the difference between the sensations on the two sides of your body.

- Let's try the left side now. You will notice how your body changes as your weight distribution changes. Pay attention to your left foot, leg and hip and then your right foot, leg and hip.

- Next, consciously and slowly raise your left leg, swing the left foot forward and place your foot on the ground. Put your weight on your left foot and take the weight off your right foot. Notice the sensation in your left foot as it carries your weight. Compare this sensation to the feeling in your right foot as you take your weight off it.

- Then, very slowly and deliberately, lift your right foot, swing it forward, place it on the ground and gently shift your weight onto it.

- Repeat this with each foot, being mindful of how each part of your body is participating in walking. Continue to notice the sensations in your feet, legs, hips, back, stomach, chest, shoulders, neck, arms and head.

- Use this mantra to help you. Say, 'Lift, swing, and place' as you repeat the process.

- Always keep your attention on the process of walking. Don't let your mind wander. If it does, it is OK. As soon as you notice that it has, though, just gently bring your attention back.

EXERCISE
Mindful driving

- When you are driving, it is easy to slip into automatic mode. How often have you driven someplace and then when you arrive you have no memory of going past certain landmarks or turning onto certain roads? Driving is often done using our subconscious ability, without much conscious thought.

- So, to improve our attention, focus and concentration, let's try driving using mindfulness. Before you get into your vehicle, look at the outside of it and notice the colour and style, whether it's clean or dirty, shiny or dull, whether it's your favourite kind of car or one you want to trade in. When you get into the car, notice how you feel as you slide into the driver's seat. Is it comfortable? Is the inside of the car neat and clean or does it need some attention later?

- As you are sitting in the driver's seat, notice the feel of the steering wheel as you grasp it with your hands and fingers. Is it warm, cold, hard, soft, smooth, slippery or sticky? Notice how the seat feels against your back and legs. Is it soft, hard or cushiony?

- Pay attention as you put the key in the ignition and start the car. Does the key slide in effortlessly or take some fiddling to get it working. How does the car sound? Does it have a nice hum or a roar? Is it running smoothly or roughly?

- Look in your mirrors and drive your car out. As you look out the window, note what you see there. Are there other cars? Is there a lot of traffic? What type of road are you driving on? Pay attention to the other cars as they come into view, as they go by in the other direction, as they travel beside

continued

you. Check the rearview and side mirrors quickly and keep your eyes on the road. Notice the cars, pedestrians, bicycles and anything else that is sharing the road or might be pulling into the road. Be alert for anything that requires you to change speed or direction.

- When you realise that your thoughts are wandering, just acknowledge it and return your attention to driving. Find some words like 'Attention, attention' to help you stay focused. Notice the traffic signals, traffic signs and the markings on the road. Notice the cars that are close to you. Notice the scenery or buildings as you drive by them.

- Keep your eyes on the road. Bring your awareness back to the feel of the steering wheel. Notice how it feels now that you have been holding on to it for a while. Repeat 'Attention, attention'. Notice how your foot feels on the accelerator. Is it easy to push down, or does it push back against your foot? If you are using a manual car, how does the clutch feel on your left foot and the accelerator on your right? How does the gearstick feel against your hand? Be conscious of changing gears. Notice how the engine sounds as you change gears. Check your mirrors and keep your eyes on the road and your awareness on the task of driving until you get to where you are going.

- Afterwards, take time to reflect on this experience and being mindful while driving. Do you remember more of the trip? Are there parts of the trip you don't remember? Is your mind clearer than usual? Do you feel like you have more energy? How far did you get before you realised you weren't thinking about driving?

- Reflect on those things that helped you keep your attention on driving.

Awareness of your body state

How can you tell if you need to calm down and decongest your brain? And as you begin to practise mindfulness, how can you monitor how well your system is dealing with the pressure around you?

Awareness of the state of your body is also known as **interoception**, a function of the insula in the brain. Our lack of interoception means we are often unaware of our own anxious state. A simple way to understand this is by taking your pulse.

EXERCISE
Feel the difference

- Try taking your pulse. There are two places where you can do this.

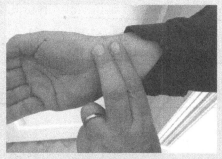

- You can place your first two fingers on your wrist (see above). If you do this very gently and don't press down, you may feel as if something is lightly tapping on your fingers. Move your fingers around a little until you can feel it.

- Next, place your first two fingers on your neck, below your jaw. Again, move them around until you feel your pulse.

continued

> - But you can make feeling your pulse even easier. Stand up. Now bend over and touch your toes ten times, fast. Now stand up again and put your finger lightly on the side of your neck.
>
> - Can you feel your pulse now? Can you count the beats? When you are angry, scared, revved up or stressed, or have been exercising, your heart beats faster, which makes it easier to feel your pulse.
>
> - Now try practising one of your mindfulness activities. Then take your pulse again. Can you feel it? It should be much slower and harder to feel.

Self-awareness

We learned in **Part 2** that our explicit memories often have deep emotional content that has been stored as implicit memories. When these are activated in the present, they can dictate our reactions and behaviour in unhelpful ways. Self-awareness is at the core of mindfulness. As our self-awareness grows, so we become more mindful.

Let's better understand this through an exercise designed to help improve our self-awareness, and 'awareness of awareness'.

EXERCISE
Implicit and explicit memories

- List three explicit memories you have from your childhood.
  1. _____

  2. _____

  3. _____

- Are they positive or negative?
- Pick one of them and describe it in detail.
  _____
  _____
  _____
  _____
  _____
  _____
  _____

- What implicit memory might be associated with this explicit memory?
- Are there any feelings that arise when you remember this memory?
- If possible, talk to someone else who was present when you had this experience. They may remember it differently, which may prompt you to reassess it.

ENERGY MANAGEMENT 1: PHYSICAL ENERGY

In **Part 1** we learned that while mindfulness is important in *relaxing* our brain, just as important is helping it work *effectively* through managing our energy levels, which will in turn help reduce stress and improve your mood.

It's hardly news that inadequate **nutrition**, **exercise** and **rest** diminish people's basic energy levels, as well as their ability to manage their emotions and focus their attention. But many people don't find ways to practise consistently healthy behaviours, given all the demands in their lives.

George worked for a large international organisation. He was significantly overweight, ate poorly, lacked a regular exercise routine, worked long hours, and typically slept no more than five to six hours a night.

After he started seeing me, George began **cardiovascular activity** at least three times a week and **strength training** at least once. (Research shows that **non-competitive, predictable and rhythmic exercise** tends to be the most effective in improving your mood and reducing your anxiety. Good exercises to pursue include hatha yoga, walking and swimming.)

George started **going to bed at a designated time and sleeping longer**. He also changed his eating habits from two big meals a day (when he normally 'gorged himself', he said) to **smaller meals and light snacks every three hours**. The aim was to help him stabilise his glucose levels over the course of the day, avoiding peaks and valleys. He lost 12 kilos in the process, and his energy levels soared. 'I used to schedule tough projects for the morning, when I knew that I would be more focused,' George told me. 'I don't have to do that any more, because I find that I'm just as focused at 5pm as 8am.'

Another key ritual George adopted was to take **brief but regular breaks** at specific intervals throughout the workday – always leaving his desk. The value of such breaks is grounded in our physiology.

We saw in **Part 2** that **ultradian rhythms** refer to 90- to 120-minute cycles during which our bodies slowly move from a state of high energy into a physiological trough. Towards the end of each cycle, the body begins to crave a period of recovery.

The signals include:

- Physical restlessness

- Yawning

- Hunger

- Difficulty concentrating

Many of us ignore these signs and keep working. The consequence is that our energy reservoir only becomes further depleted as the day wears on.

My research has found that intermittent breaks for renewal result in higher and more sustainable performance. The *length* of the breaks is less important than the *quality*. It's possible to get a great deal of recovery in a short time – as little as several minutes – if it involves a ritual that allows you to disengage from work and truly change channels. That could range from getting up to talk to a colleague about something other than work, to listening to music with headphones, to walking up and down stairs in an office building. It's even better if these breaks can involve a mindfulness activity.

While breaks are countercultural in most organisations and counterintuitive for many high achievers, their value is multifaceted.

Mike works for a South African electronics organisation. He adopted some of the same rituals as George, including a twenty-minute walk in the afternoons. Not only does it give him some exercise and a mental and emotional breather, but it's also become the time when he gets his best creative ideas. That's because when he walks he is not actively thinking, which allows the left hemisphere of his brain to give way to the right hemisphere, with its greater capacity to see the big picture and make imaginative leaps.

See if you can implement similar changes to exercise, eating and sleep in your life. You'll reap the benefits!

Some new insights into diet

Stress **actually lowers appetites for sweet food** in men and in women who are not concerned about their diet. Women who watch their diet **eat more sweets** when under stress. Stress is believed to lower some people's insulin levels, which raises the amount of blood sugar released into the bloodstream.

EXERCISE
Some fast, natural ways to relax and to go to sleep

- Spend the hour before going to bed unwinding and mentally separating yourself from the day.

- Remind yourself that it's time to become sleepy with rituals such as brushing your teeth (mindfully, if possible), locking doors, closing curtains etc. Go to bed at the same time each night.

- Limit your caffeine, nicotine and alcohol consumption.

- Exercise vigorously each day to release pent-up stress. (Wait at least two hours between doing exercise and going to sleep.)

- Before bed, do deep, slow, rhythmic breathing, as if you were already asleep.

ENERGY MANAGEMENT 2: EMOTIONAL ENERGY

When we're able to take more control of our emotions, we can improve the quality of our energy, regardless of the external pressures we're facing. To do this we must first become more aware of how we feel at various points during the day, and of the impact our emotions have on our effectiveness.

Most of us realise that we tend to perform best when we are feeling positive energy. What we find surprising is that we are *not* able to perform well when we are feeling any other way. Unfortunately, without intermittent recovery, we're not physiologically capable of sustaining highly positive emotions for long periods.

Confronted with relentless demands and unexpected challenges, we tend to slip into negative emotions – the **fight-or-flight mode** – often multiple times a day. We become irritable and impatient, or anxious and insecure. Such states of mind drain our energy and cause friction in relationships. Fight-or-flight emotions also make it impossible to think clearly, logically and reflectively. When we learn to recognise what kinds of events trigger our negative emotions, we gain greater capacity to take control of our reactions.

To break your stress cycle and make it work *for* you rather than *against* you, you need to see it as a **mind–body cycle**. Break the cycle at the **physical** level by **deep breathing**. Break the cycle at a **mental** level by changing your perception and by effective **problem-solving**.

Deep breathing

One simple but powerful mindfulness ritual for defusing negative emotions is what I call 'buying time'. **Deep breathing** is one way to do that. It counters the shallow breathing associated with a fight-or-flight type response.

Fred works for a pharmaceutical company. He had the habit of lighting up a cigarette each time something especially stressful occurred – at least two or three times a day. Otherwise he didn't smoke.

I taught him one of the breathing exercises below as an alternative, and it worked immediately. In stressful situations he found he no longer had a desire for a cigarette. It wasn't the smoking that had given him relief from the stress, but the relaxation prompted by the deep inhalation and exhalation. ✓

All the exercises you are about to learn will teach you how to breathe. Believe it or not, most of us don't know how to breathe properly – *really*.

Practise every morning for thirty days and then as needed.

> 'When we learn to recognise what kinds of events trigger our negative emotions, we gain greater capacity to take control of our reactions.'

EXERCISE
Deep abdominal breathing

Start sitting in a comfortable position, arms by your side, shoulders relaxed.

Take a deep breath. Now exhale slowly. You're probably not aware of it, but your heart has just slowed down a bit. Not to worry; it will speed up again when you inhale. This regular–irregular beat is a sign of a healthy interaction between heart and head. Inhale, beat faster. Exhale, beat slower. It's an ancient rhythm that helps your heart last a lifetime. (We saw this in **Part 2**, page 68.)

Now you've practised basic breathing, try deep abdominal breathing. In this type of breathing, your stomach goes out to inhale and is sucked in to exhale.

Exhale as completely as possible through your mouth. Feel your chest and abdominal area collapsing and falling inwards. Begin to inhale slowly through your nose, making your abdomen rise. Your chest, ribcage and shoulders should not move at all. Only your abdomen should swell as your lower lungs fill with air.

Practise fifteen to twenty cycles.

When this deep abdominal breathing feels comfortable to you, practise it as you stand and then as you walk. After that, try the exercises that follow.

EXERCISE
Body–brain breathing exercises

Stand or sit in a comfortable position, arms by your side, shoulders relaxed.

Take a deep breath out. Once your lungs are completely empty, take a deep breath in and completely fill them again. At the top of your breath count '1, 2, 3, 4, 5'. Do this three times, and with each repeat slow your breathing down. You should feel calm and relaxed.

Now complete the six exercises below in sequence, using the breathing technique you've just learned.

Exercise 1: Standing up, place one hand in the middle of your chest and the other just above your navel. Now lightly rub these two zones in a circling motion while practising the breathing technique above. Swap the positions of your hands and repeat. This exercise is **to help balance your blood pressure**.

continued

Exercise 2: While sitting in a chair, cross your right leg over your left, and hold your right toe with your right hand and your right ankle with your left hand. Continue to do this while practising your breathing through three cycles. Now change feet and hands and complete three more breathing cycles. This exercise **stimulates both sides of the cerebral cortex**.

Exercise 3: Still sitting in a chair, and with your feet flat on the floor, again commence breathing while raising your hands from your sides and placing your fingertips together as you inhale. Let your hands return to your sides as you exhale. Repeat three times. This exercise is to **balance the neural activity of the brain**.

THE NEUROSCIENCE OF MINDFULNESS

Exercise 4: To lock this balance in, extend your arms as wide as possible and picture bringing the left and right hemispheres of your brain together as you bring your hands together and clasp them as tightly as possible. Enjoy this feeling as you breathe through three cycles.

Exercise 5: Raise your arms and place two fingers from each hand above each eyebrow. As you go through three breathing cycles, place pressure on these two points – not too much pressure, but enough to feel your tension begin to disappear. The brain is now inundated with oxygen. This exercise is to **help regulate your light-headedness and start the process of reactivation out of your relaxed state**.

continued

Exercise 6: Do this **to generate *total* reactivation of the brain**. While standing, lift your left knee and touch it with your right hand. Now cross over: lift your right knee and touch it with your left hand. Continue this while going through three cycles of breathing. If you find the breathing difficult, do the physical exercise first. Once you feel confident about combining the two activities, commence the breathing sequence.

Tips on relaxing through deep breathing

- Relax with an empty stomach, as digestion can interfere.

- Program yourself ahead of time to become alert in twenty minutes. Set an alarm for twenty-five minutes to ensure you become alert when you want to.

- Relax through deep breathing for fifteen to twenty minutes several times a week.

Changing your perception

We saw in **Part 1** that the basic cause of all stress is the **perception** that you can't control a situation the way you want to. (As we'll see in **Part 5**, this has a profound effect on the immune system too.) University researchers have found that having no control can be more stressful than danger. They measured the fight-or-flight responses of racing-car drivers when they were driving at top speed, and again when they were in the pit with the crew working on the car. When do you suppose their stress was higher?

The way you view the world is, in your opinion, the correct way. All experiences and information are filtered through your personal perception, and your biases distort them to fit with your entrenched beliefs. This is called **perceptual bias**.

The ability to challenge perceptual bias and to look at situations differently is one of the most useful skills for managing stress.

Self-talk

Your perception is communicated to you through your **self-talk**. There are two ways of responding to what you perceive:

1. The **automatic response**, which includes thoughts such as:	2. The **productive response**, which includes thoughts such as:
'This is terrible!'	'Why is this so upsetting?'
'I should ...'	'What are my options?'
'Why me?'	'What information do I need?'
'I'll show them, I'll ...'	'I can control only what I do.'
Such thoughts are unproductive, and are unlikely to improve the situation.	Such thoughts are highly adaptive, and are likely to improve the situation.

If you want to change from the unproductive type of response to the more beneficial type, try thinking, 'Stop, stop!' when you realise you are thinking unproductive thoughts, and actively replacing them with problem-solving thoughts. If you want to change an outcome in your life, change unproductive thoughts into thoughts that will lead you towards the outcome you desire.

The way you explain negative occurrences can affect your future behaviour, have serious implications for your mental and physical wellbeing, and categorise you as an optimist or a pessimist.

Try to be aware of the kinds of habitual thoughts that cause you stress, and try to break those thought patterns.

Three patterns to listen for:

- **Neverending** versus **temporary**
- **Global** versus **specific**
- **Blaming self** versus **blaming other causes**

The more you explain negative things that have happened to you in neverending, global and self-blaming terms, the more you open yourself up to depression the next time something negative happens. It will also teach you to expect the same in the future.

For example, if I think my spouse left me because I'm completely unlovable, it follows that I'll expect to be completely unlovable in the future. So if a potential new partner shows me respect and love, I may sabotage the relationship because it doesn't fit with my expectations (perceptions).

Another example: a salesperson who hears herself say, 'I never reach my quotas!' could challenge this perception by saying, 'Wait a minute, how many months in the last year have I made my quota? Seven times. What was the difference between those seven months and the other five when I didn't make my quota?'

By staying with the first perception ('I never make my quotas'), she increases the likelihood that she won't make her quota next month. By challenging the perception, not only does she have a much more accurate perception, but her thoughts will also carry her in a much more productive direction.

What you expect, good or bad, is also communicated to others through a variety of verbal and non-verbal cues. These cues encourage those around you to respond accordingly, so your expectation often becomes a self-fulfilling prophecy.

If the salesperson above consistently says things such as 'I never reach my quotas', she'll give off cues supporting that assessment to others. They'll then start to behave in ways that will prevent her from meeting her quotas.

EXERCISE
Self-talk worksheet

Pull yourself out of a bad mood in a matter of minutes with this self-talk worksheet. Use the STOP method. When you think of or speak automatic, negative thoughts, say 'Stop' to yourself. Write those negative thoughts in the left-hand column overleaf. Then take a breath and deliberately change your words and sentences to create more productive thoughts. Write them in the right-hand column and then say them out loud to yourself.

continued

Automatic thoughts	Productive thoughts

Three lenses for viewing conflict

We've seen that part of changing our self-talk is avoiding self-blame. But often when we're in conflict, we cast ourselves in the role of victim, blaming *others* or *external circumstances* for our problems.

How do we change this? The most effective way is to view the situation through three lenses that are all alternatives to seeing the world from the victim perspective:

1. **The reverse lens** – asking ourselves, 'What would the other person in this conflict say, and in what ways might that be true?'

2. **The long lens** – asking ourselves, 'How will I most likely view this situation in six months' time?'

3. **The wide lens** – asking ourselves, 'Regardless of the outcomes of this issue, how can I grow and learn from it?'

All of these lenses can help us intentionally cultivate more positive emotions.

Nigella was the contact person for reporters when her organisation went through several product recalls. Over time she found her work increasingly exhausting and dispiriting. After practising the lens exercises, she began finding ways to tell herself a more positive story about her role. She realised this was an opportunity to build stronger relationships with journalists by being accessible to them, and to increase the credibility of her organisation by being frank and honest.

The power of humour

A good sense of humour can be very important in helping change your perception. It's also one of the best ways to reduce stress. It helps create a space of time between a negative event and your reaction to it. People who share humour at home and at work tend to be more productive and cooperative, and have fewer conflicts.

Laughter has a similar healing effect on your body to muscle relaxation. It may cause the release of endorphins, which are your body's natural painkiller.

Problem-solving

Often our stress centres on dealing with specific problems in our lives. Start by solving one or two problems and trying to accept that the rest of your life will be as stressful as before today.

You'll get the best results from a mixture of rational problem-solving (using the conscious cognitive processes of the brain) and intuitive problem-solving (relying on subconscious cognitive processes). Unfortunately, society overvalues the rational approach and undervalues the intuitive, even though it's common to hear managing directors say they depend upon their 'hunches' and 'feelings' in decision-making.

EXERCISE
Mindful problem-solving

Here are five steps that will greatly increase your chances of solving your problems:

1. Use the **problem-solving worksheet** on page 114.

2. Set a **deadline** for accomplishing your problem-solving goal. This deadline can be flexible.

continued

> 3. **Tell someone else** what your goal is. Make sure this is someone who'll be supportive.
>
> 4. **Imagine** your desired outcome. Visualise yourself accomplishing your goal, step by step, and on time. Imagine the positive emotions you'll feel as you reach it. Visualise this scene three or four times each day, for a minute or two on each occasion. The best times for this are before you go to bed at night, when you wake up in the morning, or after doing a relaxation exercise (though any time of the day would be beneficial).
>
> 5. Make clear, concise **affirmations** that will lead you towards your goal. They should be positively stated, and in the present tense. For example: 'I'm calm and relaxed.' (A negatively phrased statement would be, 'I'm not defensive any more.') Make two or three statements for each goal, and say each of them ten or more times daily.

Anticipate and prepare for times when you know you're going to be under stress. Use imagination and affirmations to rehearse handling your situation and reaching your desired outcome. Three weeks or more of imagining and affirmations produce the best results.

For example: the board of directors is flying in for a visit. Normally when they're around, you feel nervous and tongue-tied. You believe they ignore your opinions.

Imagine yourself relaxed in their presence, free of muscle tension and that nervous laugh you get when they're around. See yourself speaking intelligently and being taken seriously.

Make **affirmations** like: 'I'm calm and relaxed in the board's presence.' 'The directors are listening to me and are impressed with what I have to say.'

You'll find that imagination and affirmations are also useful for getting your confidence back after you've been in a stressful situation. For example: a customer has just vented her anger at you, and you handled it poorly, enraging her even more. Most people will rehash this kind of stressful situation in their minds over and over again – which only increases the likelihood that they'll handle similar situations poorly in the future.

Instead, **imagine** handling the customer effectively, using your listening skills to calm her down; visualise her calming down, and imagine yourself directing her from anger to problem-solving. Make affirmations like: 'I'm calming upset customers effectively, using listening skills.' 'I remain calm in the presence of other people's anger.'

EXERCISE
Use a problem-solving worksheet

Use the questions opposite to analyse and allay stress.

1. Ask yourself whether there is a reason for the stress you are feeling. Make sure you go beyond the symptoms, enabling you to describe and analyse the underlying problem. Ask questions such as: 'Why is it a problem?' 'Whose problem is it?' 'What should happen next?'

2. Decide what your goals are – that is, what you want (or need) in relation to the problem – and list your options in the table opposite.

continued

What I want/need more of:	What I want/need less of:

3. Having decided on your goal, now create an action plan that will take you towards it, using the table below:

How to obtain what I want/need more of:	How to avoid what I want/need less of:

4. Evaluate and adjust your plan.

The tensions of our lives will always be there, but by refuelling our emotional energy – both physically and mentally – we'll gain the resilience to stop them from undermining our effectiveness.

> *'The ability to challenge perceptual bias and to look at situations differently is one of the most useful skills for managing stress.'*

ENERGY MANAGEMENT 3: MENTAL ENERGY

Many of us view multitasking as a necessity in the face of all the demands we juggle each day. But it actually undermines our productivity.

The interruptions that technology has introduced to our lives are relentless. Do a simple check for yourself. Pick any one-hour slot in your day and count the number of interruptions that occur. Notice how much you struggle to concentrate as a result. It'll be very enlightening.

Distractions are costly. A temporary shift in attention from one task to another – stopping to answer an email or take a phone call, for instance – increases the amount of time necessary to finish the primary task by as much as 25 per cent: a phenomenon known as **switching time**. It's far more efficient to **focus fully** for **90 to 120 minutes**, take a **true break**, then focus fully on your next activity. These work periods are known as **ultradian sprints**. Such focused thinking is another effective form of mindfulness.

You can also create rituals to reduce distractions. Don, an employee at an American bank, designed two rituals to help him focus better throughout the day. The first one was leaving his desk and going to a conference room, away from phone and email, whenever he had a task that required concentration. He now finishes reports in a third of the time.

Don built his second ritual around meetings with those who reported to him. Previously, he would answer his phone whenever it rang during these meetings. As a consequence, the meetings he scheduled for an hour often stretched to two, and he rarely gave anyone his full attention.

Now he lets his phone go to voicemail so he can focus completely on the person in front of him. He answers the accumulated voicemail messages when he has downtime between meetings.

Develop your own rituals

Most of us who work in an office reply to our emails constantly throughout the day, whenever we hear the 'ping'. Many of us are glued to our smartphones. We need to create a ritual of **checking our devices just twice a day** – perhaps at 10.30am and 2.30pm.

Where we previously couldn't keep up with all our messages, we should now be able to quickly clear our phone or email inbox each time we check it. This is the reward of fully focusing our attention on our messages for forty-five minutes at a time. It also resets the expectations of all the people we communicate with. Tell your colleagues that if there are any emergencies, they should call you and that you guarantee the phone will be picked up. You will be surprised at how few calls you get.

Another ritual is to focus on activities that have the most long-term impact. Unless we intentionally schedule time for more challenging work, we will probably either rush through it at the last minute, or not get to it at all.

Perhaps one of the most effective focus rituals we can use is to **identify each night the most important challenge**

for the next day, and make it the very first priority in the morning. Instead of answering messages first thing in the morning, try concentrating for the first hour of every day on your most important and challenging task. By 10am you'll already feel you've had a productive day.

Focused thinking is a way of actively concentrating your mind on your work. When it wanders, force it back to the topic at hand. Don't give it an inch. You may find this focused thinking difficult at first, but your mind will soon get used to it.

Up to 25 per cent more productivity by putting a few simple changes in place? Who *wouldn't* give that a go?

ENERGY MANAGEMENT 4: MINDFUL ENERGY

We tap into the energy of our human spirit when our everyday work and activities are consistent with what we value most, and with what gives us a sense of meaning and purpose. If what we are doing really matters to us, we typically feel more positive energy, focus better and show greater perseverance.

Regrettably, the high demands and fast pace of modern life don't leave much time to pay attention to these issues. Many of us don't even recognise meaning and purpose as potential sources of energy.

Only when you've experienced the value of the rituals from the other energy dimensions do you start to see that being attentive to your own deeper needs has a dramatic influence on your effectiveness and satisfaction in both work and life.

Sometimes just having the opportunity to ask yourself questions about what really matters to you is both illuminating and energising. It's important to be a little introspective and think about what you want your life's legacy to be. You don't want to be remembered as nothing more than that crazy manager who worked very long hours and whose team was miserable. You don't want to be the father or mother who comes late to the kids' show, sits at the back and is looking at your smartphone more than at the concert.

To gain access to the energy of mindfulness you need to clarify priorities and establish rituals in three areas:

1. Do what you **do best** and **enjoy most**.

2. Consciously allocate time and energy to **the areas of your life you consider most important** – work, family, health, service to others and whatever else you feel is significant.

3. Live your **core values** in your **daily behaviour**.

1. What you do best and enjoy most

It's important to realise that these aren't necessarily going to be the same thing. You may be very good at something and get lots of positive feedback about it but not truly enjoy it. Conversely, you could love doing something but have no gift for it, so that it requires much more energy than it makes sense to invest.

For example, Julia, a senior manager, realised that one of the activities she least liked was reading and summarising detailed sales reports, while one of her favourites was brainstorming new strategies. Julia found someone on her team who loved immersing himself in numbers and delegated the sales reporting to him, settling for brief oral summaries each day. Julia then began scheduling a free-form ninety-minute strategy session every other week with the most creative people on her team.

So, what are your areas of strength? Think of two experiences in the past few months when you found yourself in the 'sweet spot' or the 'zone' – feeling effective, effortlessly absorbed, inspired and fulfilled.

Now deconstruct these experiences to understand precisely what made them so positive, and what specific talents you were drawing on. Is it being in charge of a project that invigorates you? Or taking part in something creative? Or is it using a skill that comes to you easily and feels good to use?

Finally, you'll need to establish a ritual that will encourage you to do more of exactly that kind of activity.

2. Areas of life that are most important

Just as what we do best and enjoy most aren't always the same, there's often a divide between **what we say is important to us** and **what is actually important**. Again, rituals can help close this gap.

When Alice thought hard about her personal priorities, she realised that spending time with her family was what mattered most to her, but it was often squeezed out of her day. So she started a ritual where she switches off for at least three hours every evening when she gets home, so she can focus on her family.

In the evenings she used to talk on her phone all the way to her front door. She now puts the phone away twenty minutes before she reaches home. She uses this time to relax and unwind, so that when she does arrive home she's less preoccupied with work and more available to her family.

3. Living your core values in your daily behaviour

This is a challenge for many of us. We're living at such a furious pace that we rarely stop to ask ourselves what we stand for and who we want to be. As a consequence, we let external demands dictate our actions.

Rather than explicitly trying to define your values, try asking yourself, 'What are the qualities that I find most offputting in others?' By describing what you *can't stand*, you'll start to realise what you *stand for*. If you're offended by stinginess, for example, generosity is probably one of your key values. If you're especially put off by rudeness, it's likely that <u>consideration for others is a high value for you</u>.

As in the other categories, establishing rituals can help bridge the gap between values you aspire to and how you currently behave. If you discover that considering others is a key value, but you are perpetually late for meetings, your ritual might be to <u>end your previous task five minutes earlier</u> than usual and <u>intentionally show up for meetings five minutes early</u>.

> *'We're living at such a furious pace that we rarely stop to ask ourselves what we stand for and who we want to be.'*

ENERGY MANAGEMENT SUMMARY

Addressing the four energy management categories discussed will help you gain a greater sense of satisfaction and wellbeing in all areas of life. Those feelings are a source of positive energy in their own right, and will reinforce your desire to persist in improving your management of other energy dimensions as well.

Here's a summary you can use in putting together a personal plan:

1. Physical energy

- Reduce stress by doing cardiovascular **exercise** at least three times a week and strength training at least once.

- Enhance your **sleep** by setting an earlier bedtime, relaxing before bed and reducing caffeine, nicotine and alcohol use.

- Improve your **diet** by eating small meals and light snacks every three hours.

- Increase your **awareness** of signs of flagging energy, including restlessness, yawning, hunger and difficulty concentrating.

- Take brief but regular **breaks**, away from your desk or workplace, at 90- to 120-minute intervals throughout the day.

2. Emotional energy

- Defuse negative emotions by practising deep **breathing**.

- Change your **perception** of stressful and upsetting situations by using productive self-talk and seeing conflict through new lenses.

- Use a mixture of rational and intuitive **problem-solving**; imagine your desired outcome and make positive affirmations that will lead you towards it.

3. Mental energy

- **Focus** fully on one activity for 90 to 120 minutes, then take a true break.

- Respond to **messages** only at designated times during the day.

- Every night, identify the most important **challenge** for the following day. Then make completing that job your first priority the next morning.

4. Mindful energy

- Find your **'sweet spot'** – what you do best and what you enjoy most.

- Make the most important areas of your life your top **priorities**.

- Live your core **values** in your daily behaviour.

Putting together your plan

To ensure you create an environment in which you continue to practise these rituals, you'll need to put together a spreadsheet for yourself. An example of such a spreadsheet is below. Put your own together and aim to make it an everyday part of your life from this point on.

DATE:

1. Physical energy

	Current	Goal	Month 1	Month 2
Exercise				
Sleep				
Diet				
Awareness				
Breaks				

2. Emotional energy

	Current	Goal	Month 1	Month 2
Breathing				
Perception				
Problem-solving				

3. Mental energy

	Current	Goal	Month 1	Month 2
Focus				
Messages				
Challenge				

4. Mindful energy

	Current	Goal	Month 1	Month 2
Sweet spot				
Priorities				
Values				

Set up your own rituals and check them off over a couple of months (or longer) until they become **habits**. They're the key to **managing your energy, NOT your time.**

PART 4

FROM MINDFULNESS TO WELLNESS: THE MIND–BODY CONNECTION

AN INTRODUCTION TO THE MIND–BODY CONNECTION

Back in the early 1980s, there were three of us doing our doctorates while working together in a hospital in Melbourne. Two of us went on to work in medicine, and one of us (me) went on to specialise in psychology and cognitive neuroscience.

My two colleagues went over to the USA to work in the new neuroscience laboratories whose work with Canadian psychiatrist Norman Doidge, author of the bestselling *The Brain That Changes Itself*, was just beginning. The three of us would catch up for a beer every now and again, when my colleagues were back in Australia, and the challenge they always put to me, as medicos and surgeons, was: 'Well, you know, Stan, the day you can prove to us that a thought can convert into something physical, and show us how that occurs, is the day we'll believe you. We might even buy you a beer.'

Amyloid protein

What they were asking for was the essence of **the mind–body connection (MBC for short)** and subsequently its role as **a foundation for mindfulness**. That was almost forty years ago, and it has really only been in the last few years that the big breakthrough has occurred. And that big breakthrough was the discovery of the **amyloid protein** (that scum around the bath that I mentioned in **Part 1**). My two colleagues now owe me an endless supply of beer!

Now, amyloid protein is exactly that – it's a protein, it's matter. And it was the process by which the amyloid proteins build up in the brain that was the big discovery. Using MRI technology, scientists were actually able to track the chain of events. They found clear evidence that when we have a thought that is stressful, or when an event is stressful – and remember the brain doesn't differentiate between the two – this leads to the buildup of these harmful proteins in the brain. We knew the mind–body connection existed, but this was the first time that we were actually able to show it and prove it, in a physiological way.

Almost forty years ago, my two colleagues had asked me: 'You know, Stan, you know, how can a thought lead to a heart attack? How can a thought lead to a brain aneurism?' Discovering amyloid protein gave us the answer. If I'm stressed or anxious, and my thinking continues to keep me in that state, and if the cortisol stays there, eventually it forms this protein. And the protein literally builds up in the blood vessels, in both the brain and the heart, resulting in heart attacks and aneurisms. A crystal-clear example of the mind–body connection.

Eastern medicine

This connection between the mind and the body has been well known for centuries in Eastern medicine, where there is not the same emphasis on rigorous scientific evidence as we have in the West. At one extreme, there's a suggestion that Buddhist monks who practise extraordinary levels of meditation can actually communicate with each other through thought alone. Recently there was a news story of a monk who was in such a deep state of hypnosis that he appeared dead; he was buried for three months before being dug up and found to be alive!

For many years, there have also been doctors and others in the West talking about the powers of meditation – the powers of being able to focus on your cancers, for example, and shrink them. All of a sudden the scientists are saying, 'We might not quite believe *that*, but we possibly could slow things down, or prevent them from starting in the first place.' We didn't understand the mechanics of it previously, but we do now, thanks to the discovery of the amyloid protein. Today, for example, there are doctors and dentists, particularly in the United Kingdom, who don't use anaesthetics, but hypnosis and meditation.

There's a lot of stuff out there that comes across my desk and I think, 'Really? This is far-fetched.' But I once thought that about a lot of things I now accept as scientific fact.

The placebo effect

The placebo effect occurs where the benefits of a treatment come purely from belief in the treatment, not from the treatment itself. And there are many studies showing that placebos outperform actual depression drugs.

This isn't to say that medicines aren't important, but in some cases where doctors apply the old-fashioned treatment model, perhaps they don't understand the great power of just being positive, of believing in something *beyond* the drugs. The mind's connection to the body is simply not yet well enough understood.

Your body

The biggest problem people have in understanding and accepting MBC is that they can't see how 'thoughts' can lead to actual changes in their body. This is probably because most of us don't know very much about our body itself.

Your body is one of the most extraordinary creations on this planet. Your brain, by itself, is so complex that scientists can't even imagine ever building a computer that can do what it does (yet).

Because the body is so complicated, the professionals who study it – doctors, biologists, immunologists, to name a few – have developed a huge range of specialist terms and names for the various parts and systems within it. It's this specialist terminology that discourages a lot of us from discovering more about our own bodies.

Understanding a little about MBC – both the technical and the practical sides – can break down some of those barriers and give us the confidence to start applying MBC principles to our own lives. It's like an advanced form of mindfulness.

MBC is the foundation of mindfulness. This foundation needs to be understood and then maintained to ensure the integrity of the whole system. The mindfulness techniques we use in this book allow us to improve the balance (homeostasis) of our physical and mental functioning.

The three parts of MBC

M = mind Mental/emotional factors: emotions, thoughts, beliefs, attitudes, coping styles.

B = body The brain, the nervous system and the endocrine system. The nervous system can be broken down into the autonomic, sympathetic and parasympathetic nervous systems. The endocrine system is the system of glands that produce and control various hormones.

C = connection The immune system, the system of cells and chemicals in your body largely responsible for connecting, detecting and dealing with bacteria, viruses, cancer cells, parasites and any other living or non-living particles that can lead to illness.

We'll look at each of these in greater detail in **Part 5**.

'The big breakthrough in MBC was the discovery of the amyloid protein. Stress leads to the build-up of this harmful protein in the brain.'

MBC RESEARCH

Research into MBC connects medical science and health psychology. It looks at the relationship between your psychology and your immune system. It examines how your coping styles, emotions and attitudes and the events of your life affect your body's nervous system, hormones and immune system, and ultimately your health.

Unlike other types of psychology, however, MBC is only interested in measurable changes in the body. Success or failure in MBC concerns *physical* measurements, not *emotional* measurements.

Here are just a few of the areas that MBC research has examined:

- What does stress actually do to your body, in terms of hormones, cells and your overall health?

- Do people who get cancer, heart attacks or other illnesses have different attitudes and coping styles from those who do not? If so, how might their attitudes and coping styles have contributed to the development of the illness?

- Do people who recover from serious illness, or who survive longer, have specific coping styles that make them different from other people? If so, how do these coping styles affect their body's ability to deal with the illness?

- Can specific psychological methods for releasing emotions – such as mindfulness – actually produce changes in the immune system? Can these be used to prevent and even treat people with illnesses such as cancer and HIV/AIDS?

- Do major events in our lives – moving home, losing a job, divorce, death in the family – affect our health? If so, how?

- Why is it that, when people suffering from the same illness receive the same medical treatment and follow the same basic diet and exercise routines, only some pull through, while others don't?

Strictly speaking, MBC is an area of research. But when we start taking the results of this research and apply it to improving people's health, or preventing disease, we call it **Applied MBC**. All the mindfulness techniques you've learned so far in this book could be part of an Applied MBC approach, along with some more specific strategies we'll look at in **Part 6**.

So far, MBC has produced some pretty interesting – and useful – results. The research in this area is going ballistic, as scientists look at the role of stress hormones, and the resultant protein, in any number of medical conditions. They still have little idea of the full extent of cortisol's effects, especially because it appears that it can stay in the body's systems for a long time. But if a thought can *lead* to physical outcome, surely a thought can equally *remove* that outcome. That's the basic principle of MBC.

Immune system illnesses

Immunology seems to be the next big MBC frontier. We don't know yet just how invasive chronic stress is, but we now have enough knowledge to confidently state that it has a very negative impact on our overall immunology. There is a lot of data saying that if you can relax your system you will improve your immunology. (We'll look at this more in **Part 5**.)

How many people understand that when they're run down, they catch things? Their whole defence system is being trained on combating the effects of stress, so that they become open to other illnesses, such as colds. If you think about the effects of cortisol that we can see, imagine what it's also doing to our cellular structures.

So far, the conditions that have responded best to Applied MBC are two groups of immune-related diseases. Whether the cause was a virus or bacteria, in these diseases the immune system is either over-active or under-active – either acting against itself, or failing to act where it should.

1. Autoimmune diseases

Autoimmune diseases – such as **rheumatoid arthritis, systemic lupus erythematosis**, and **multiple sclerosis (MS)** – occur when the immune system mistakenly attacks 'normal' parts of the body. In **rheumatoid arthritis** and **systemic lupus erythematosis**, the immune system destroys the tissue in the joints, while in **MS** it attacks the 'insulation' around the nerves – myelin sheathing – causing 'short-circuiting' of nerve impulses to the limbs. Other, lesser-known diseases in this category are **progressive systemic sclerosis, giant cell arteritis, polymyalgia rheumatica, polymyositis** and **polyarteritis nodosa**. Another way of viewing these diseases is to see them as the result of an over-active immune system.

2. Suppressed immune system diseases

These include **cancer, HIV/AIDS** and **chronic fatigue syndrome (CFS)**. In all these diseases, the immune system is not reacting as it should – for some reason or other, it is apathetic and lazy.

For example, **cancer** cells develop as a matter of course in every person. A cancerous cell is a cell that refuses to die and just carries on growing. These cells accumulate and use up blood and resources, thus starving the adjacent tissue. They also put pressure on neighbouring organs. The usual job of our white blood cells is to identify cancerous cells and destroy them. However, when the immune system is suppressed, this does not occur and tumours develop.

In **HIV/AIDS**, the virus attacks and occupies the very cell responsible for sending out the chemical message that something is wrong. Other cells, such as natural killer cells, do not require this prompting and should clean up the virus, but in most cases they don't. Once again, the immune system doesn't respond as it should, and in about 95 per cent of cases, this failure gradually weakens the immune system, allowing bacteria and viruses to wreak havoc with little resistance. (The other 5 per cent of people infected with HIV *don't* develop AIDS, and their immune systems stay strong; these people are called 'long-term non-progressors'.)

In people with **CFS**, the immune system crashes, for reasons that are still unclear. The crash appears to be quite sudden, and the system remains weak for quite some time, even years. One theory is that this is because the person's immune system didn't recover properly from a previous viral infection (specifically, Epstein-Barr virus). Another theory is that it is caused by depression. The disease is not fully understood, and some do not accept it as an identifiable illness. Women seem to be affected far more than men, and it tends to be more common in people between the ages of twenty and thirty. It's not uncommon to find that people with CFS were high-powered, go-go-go people, whose world crashed due to some failure. Too much adrenaline, for too long.

Type 2 diabetes

Another disease that is becoming a big focus of MBC research is **type 2 diabetes**, which tends to be lifestyle-related. As I mentioned back in **Part 1**, on the whole our lifestyles have got healthier, despite what the media tell us. But for all of our improved eating, exercising and other habits, a lot more of us now have type 2 diabetes, or are on the cusp of developing it. The only thing we've got worse at is stress.

The research continues

By no means have we discovered *all* of the health problems that can be at least partly attributed to the mind, and we wouldn't want to attribute *everything* to it, but we can confidently state that it's a player in a lot of medical conditions. Only time will tell us just how many.

SOME WORDS OF CAUTION

What MBC research does and doesn't claim

In my introduction to MBC, I mentioned that the discovery of amyloid protein explained how a thought can lead to a heart attack. But it would be silly to suggest that psychological factors *caused* you to be infected with HIV, or *caused* you to have cancer or some other disease. Germs and cancer cells exist, regardless of what you think and feel.

MBC research does not say that psychological factors *cause* disease. All it says is that certain psychological (and social) factors contribute to the *development* of some diseases, and also to the *healing* of those diseases.

There are a number of ways that psychological (and social) factors *can* contribute to illness:

1. Your psychology has an effect on whether you will say yes or no to taking the risk of being exposed to certain types of germs. Being infected with HIV could be an example of this.

2. How you cope with life in general also affects your immune system, to the degree that your body will

fight (or fail to fight) germs when you are first exposed to them.

3. Once the germ (or cancer cell) has started to affect your body, psychological factors make a difference in how well (or not) your body deals with it.

4. Genetic predispositions to certain illnesses can be either lessened or strengthened by how you cope with life.

Not every illness has a psychological cause or component. Sometimes you just ate the wrong things, or breathed dirty air, or something went wrong in your physical environment.

What MBC is and isn't

MBC and medical treatment

Applied MBC has never been – and should never be – viewed as a stand-alone intervention. It should *always* be carried out alongside proper medical treatment, nutritional assistance, exercise, and whatever else is appropriate for a particular illness.

Mindfulness and other MBC methods are not in any way a replacement for standard medical treatment. MBC is not an alternative to standard medical treatment – it is a complementary approach that works best alongside it.

No one is talking about throwing out hundreds of years of outstanding medical science. While using MBC, it's vital to ensure that your health is medically monitored and treated. How are you going to know whether your disease-fighting T cells are increasing if you don't get a medical blood test?

It might also be a good idea to consult a professional therapist to help with any significant psychological issues. You could even ask them to help you work through the issues and methods in this book.

So, just to make it clear: following the proper medical instructions is absolutely essential. What MBC does is to *improve* on the effects of medical science. At the very least, MBC can support recovery by strengthening the immune system. If you're sick and worried about the consequences of your illness, your body is already stressed and your immune system is weakened. But if you can actually get yourself relaxed, you'll turn the cortisol in your system from a horror drug into the wonder drug it's supposed to be. Just a little bit of mindfulness, for instance, can get rid of the cortisol and provide a real boost to your immune system.

MBC and alternative therapies

MBC is not spiritual healing. It is not about 'mind over matter'. It is not 'new age' either.

It's understandable that practitioners of various alternative healing methods feel somewhat supported by MBC research. But rarely do these people properly understand what MBC is, nor do they understand the research or the biological pathways involved.

I don't mean to criticise non-medical modes of healing, or to comment on religious belief, faith, chakras, auras, the existence of the soul or universal love. The reason is simple: until any of these can be **scientifically demonstrated** to have an impact upon the immune system, and their biological mechanisms can be explained, **these methods of healing cannot refer to themselves as MBC-related**.

This does *not* mean that these phenomena do not exist, or that they are somehow less real or valid. It simply means that they are not MBC.

And yes, sometimes we do things because they work, even when we don't know *how* they work. Every day, millions of people take aspirin for pain, even though we still don't know how it works!

And, crazy as it might sound, **you don't even have to believe in MBC to benefit from it**. You do, however, have to do some work – and that's what I'll be focusing on in **Part 6**. First, though, we need to look at the different components of MBC in more detail.

'MBC is not an alternative to standard medical treatment – it is a complementary approach that works best alongside it.'

PART 5

HOW MBC WORKS

THE 'M' PART OF MBC: MIND

The 'mind' part of the MBC puzzle includes thoughts, attitudes and beliefs, as well as feelings such as guilt, anger, fear, depression, anxiety, joy and excitement. In addition, there are your perceptions – the ways in which you view or interpret the world and what is happening – and all your memories from birth onwards. (Some would even say that you have memories from conception and the womb.) Then there are your attitudes, coping styles, skills, beliefs, values and morals.

Hopefully what you've read in this book so far has convinced you that what goes on in your mind has a vital impact on your health. The following psychological factors have been clearly demonstrated to weaken the immune system:

- Bereavement
- Low marriage quality, divorce and separation
- Miscarriage or failure to conceive
- Unemployment

- Retirement
- Taking care of someone with a serious disease
- Living in a dangerous place
- Exam-related distress
- Absence of social support
- Loneliness
- Anxiety and depression
- Fear
- Low self-worth
- Suppression of anger and other emotions
- Using denial to cope with problems
- Stressed power syndrome (see **Parts 1** and **6**)
- Insomnia

But it's all well and good discussing whether psychological factors can *negatively* affect the immune system. A more important question is: 'Can I *positively* affect my immune system by using psychological methods?' The answer is a definite 'Yes'. There are literally hundreds of studies demonstrating that the immune system can be strengthened through psychological intervention.

One example is a study of fifty healthy college students that examined the effects of confronting painful past experiences. Half the students were asked to write about some superficial topic (e.g. how to decide which vegetables to buy at the supermarket), while the other half wrote about a personally painful past experience (e.g. a relationship break-up, the loss of a loved one, a serious accident, or failing an important exam). All students wrote for twenty minutes per day on four consecutive days.

The students who wrote about painful upsets in their past had stronger immune responses immediately afterwards, and in the longer term, paid fewer visits to the college health clinic. Three months later, these students were significantly happier than the others. It was also found that students who had previously held back from talking to others about their painful experiences benefited more than those who had previously discussed their experiences with someone else.

The researchers concluded that suppressing painful experiences is stressful. Talking or writing about such events – getting them off your chest – removes some of that stress, and also benefits the immune system.

One important point is that the students didn't require any medical therapy; they simply wrote about their experiences. In other words, as we've already seen with mindfulness, Applied MBC methods can be used by ordinary people, at home.

THE 'B' PART OF MBC: BODY

When people start to learn about MBC, most can relate to the 'mind' part. But when they try to look into the 'body' side of things – the brain, nervous system and endocrine system, and the various chemicals the brain releases that influence the immune system – they get confused.

The rest of **Part 5** aims to explain, as simply as possible, the why and how of this interaction between your thoughts, feelings and attitudes and your health.

Remember: you *don't* have to understand how the various hormones and brain and nerve pathways do what they do. This understanding is useful when you're trying to explain MBC to someone else, but you don't need it to benefit from MBC yourself.

Have you ever wondered how you think about lifting your finger, then your finger moves? How does a thought travel through your system and instruct your body to do things?

The most obvious mechanisms behind this are the **brain** and **nervous system**, which together make up a highly intricate system of 'wiring'. Electrical signals are sent back and forth between your brain and your body through nerve 'wires', with your brain acting as the main 'switchboard'. So, for example, you think (unconsciously), 'Finger – lift!' and the brain translates this instruction into an electrical impulse that races through your nerves to your finger. The finger then obeys the instruction, and causes the finger muscles to lift.

Most people believe that the brain–nerve system is the primary control mechanism in the human body. In reality, there are *two* equally powerful control systems: the **brain and nervous system**, and the **endocrine system**.

The brain and nervous system

Your brain is connected to the rest of your body by means of the spinal cord, which runs through the centre of your spine, and from which nerves extend into the rest of the body. Together, the brain and the spinal cord are called the **central nervous system (CNS)**. The major nerve system that connects the spinal cord to the rest of your body, especially your organs and internal muscles, is called the **autonomic nervous system (ANS)**.

The autonomic nervous system has two types of nerves: the **sympathetic nervous system (SNS**, which, for example, causes your heart to beat faster), and the **parasympathetic nervous system (PNS**, which, for example, slows your heart rate down). (We met these when we looked at the fight-or-flight system back in **Part 1**.)

Let's now look more closely at the ANS, SNS and PNS in turn.

The autonomic nervous system (ANS)

The ANS functions continually, whether you are awake, asleep or even under a general anaesthetic. It may be the only part of the nervous system keeping you alive when all other parts of the brain have shut down (e.g. in a coma). It controls all your non-skeletal muscles:

- Your heart
- Your lungs
- Your kidneys
- Your gastrointestinal tract (stomach)
- Your glands

Most functions of the ANS happen without conscious control, but some are able to be controlled through activities like deep breathing and mindfulness.

The sympathetic nervous system (SNS)

As we saw in **Part 1**, a threat to your survival or a stressful situation activates the branch of the autonomic nervous system known as the sympathetic nervous system (SNS). It provides the body with a sudden burst of energy. As we've seen, it was originally a survival mechanism, but in our modern world it is often set off inappropriately by situations we *perceive* as threatening or stressful.

The SNS sends a message to the adrenal glands to secrete the hormones adrenaline and noradrenaline. These two are directed at specific cells in particular muscles, organs and glands. As a result, blood pressure, respiration rate and heart rate all increase to maximise the amount of oxygen supplied to the muscles that have to do the work.

The parasympathetic nervous system (PNS)

This system serves two main functions:

1. Once the threat (or perceived threat) is over, it returns the body to a calm state by reversing the changes brought about by the SNS.

2. It minimises energy use and keeps the internal systems

of the body constantly regulated. This process is known as **homeostasis** (which we looked at in **Part 2**).

The PNS is the dominant system because it is involved with normal everyday functioning: regulating food digestion, waste elimination and tear production.

Now let's look at the body's *second* major control system.

The endocrine system

Throughout your body are a number of organs called **endocrine glands**, whose primary function is to produce **hormones**. These hormones have the ability to control other cells in your body.

For example, the **pancreas**, an endocrine gland located beneath your stomach, secretes the hormones insulin and glucagon, which control the level of blood sugar in your body, which in turn affects your metabolism and energy levels. Another well-known gland is the **thyroid gland**, located in your throat, which also releases hormones that affect the metabolic rates of your cells. Two other sets of glands, the **ovaries** in women and the **testes** in men, release sex hormones such as oestrogen and testosterone.

Two very important glands – in terms of MBC – are the **adrenal glands**, which are located on top of the kidneys. These glands have two parts: the **adrenal medulla**, which secretes adrenaline hormones, and the **adrenal cortex**, which secretes cortisol hormones. We've met these two hormones many times before in this book, and seen how important they are in understanding the effects of stress.

But the most important gland for the purposes of MBC is the **pituitary gland** – the gland that controls all of the other glands. It's located inside the brain, and is connected to the **hypothalamus** by means of a tiny vessel called the hypophyseal (or pituitary) stalk. The hypothalamus is sometimes referred to as 'the seat of emotions'. Directly or indirectly, all emotions and thoughts pass through, and are processed by, this part of the brain.

This connection between the pituitary gland and the hypothalamus is the most obvious of the connections between what happens in the mind and what happens in the body.

But there's a *third* system in the body that we need to take into account in MBC: the immune system. That's the 'C' part of MBC.

THE 'C' PART OF MBC: CONNECTION

Connection: the final piece of the MBC puzzle. This occurs through the **immune system**.

Your body is *designed* to deal with disease. The important question is why it sometimes doesn't succeed in this task. Often, the problem is not the germs or cancer cells but the fact that your immune system isn't dealing with them properly. This is why we speak of 'strengthening the immune system'.

For example: why does the body sometimes attack itself? This is the reason for the *autoimmune system illnesses* we looked at in **Part 4**. And why does the body not send the necessary fighter cells to cancer tumours, even when those cells exist? This is an example of a *suppressed immune system illness* (which we also looked at in **Part 4**).

The immune system has been described as your body's way of telling the difference between self and non-self. It's the body's way of protecting itself, and attacking anything that is *not* itself. There's a fascinating connection between your psychological sense of self, and your body's sense of self. When the psychological aspect is not balanced or strong, this has a direct effect upon the physical aspect.

The immune system

Most people understand the immune system as the cells that attack cancer cells, parasites, bacteria and viruses. But the immune system is much more complicated than this. It involves each and every defence method your body has against germs and poisons – including the wax in your ears, the saliva in your mouth and the hairs in your nose, plus all the cells in your blood, various glands, neurotransmitters and more.

The non-specific immune defence system

Imagine that your body is a city full of people. Around this city is a huge wall – your skin – that prevents most intruders – bacteria, viruses and parasites – from getting inside your body. There are also several gates into this city: your ears, mouth and nose. All of these gates are protected as well. Your ears have wax, your eyes have tears, and your nose has hairs and mucus, all of which prevent intruders from entering your body. If the intruders get past these body defences, the chemicals in your saliva or the acid in your stomach will usually kill them.

All these parts of your body, and the substances they produce, form an important component of your immune system. Collectively, they're called the **non-specific immune defence system**. More specifically they include the following protective mechanisms:

- Coughing germs out
- Sneezing germs out
- Wafting of cilia (hair-like things) lining the trachea and bronchi (throat and lungs)

Secreted substances:

- Wax in your ears
- Enzymes in your tears
- Chemicals in your saliva
- Hydrochloric acid in your stomach

Cells that attack as soon as something enters the body, including:

- Macrophages – a type of cell that consumes foreign microbes
- Neutrophils – white blood cells that combat infection

Circulating substances:

- The complement system – a group of proteins in the blood that help eradicate infectious microorganisms
- Interferons – proteins released in response to the presence of bacteria and viruses

Special cells:

- Natural killer cells – white blood cells that kill cancerous cells and cells infected with a virus

Specific immune defence system

It's only when the intruders get beyond the barriers of the non-specific immune defence system that the second line of defence kicks in, namely the cells in your blood. They're like the forces waiting inside the walls of the city. This is called your **specific immune defence system**, and it has two main components: the **humoral system** (the police) and the **cell-mediated system** (the army).

1. The humoral system:

- Deals mainly with bacteria and parasites
- Consists of B cells (B lymphocytes) and plasma cells
- B cells identify the germ (antigen), and produce an identikit (memory) of it
- Antibodies (immunoglobulins) are produced by plasma cells, based on the B cells' identikit; these antibodies 'arrest' and lock onto the germ, so that other cells (e.g. macrophages) will recognise it as an invader and kill it

2. The cell-mediated system:

- Specialises in germs that are already inside tissue and cells (e.g. viruses and tumours), and works well for bacterial infections
- Consists of T cells (T lymphocytes), which recognise the germ in the cell or tissue, attach to it and kill it

All the cells involved in immune defence are **white blood cells**. They form part of the non-specific immune defence system (macrophages and neutrophils), the humoral system (B cells and plasma cells) and the cell-mediated system (T cells). **Red blood cells** transport oxygen through the blood within molecules of haemoglobin, and are not directly involved in the immune system.

How cells talk to each other

So how do all these tiny cells communicate with each other? How does the liver know when to release glucose? How does the heart know when to pump more blood than normal and when to slow down?

There are two forms of communication: via the **nervous system**, and via the **chemical system** (including hormones).

Imagine the complicated process that must occur when one cell discovers a virus or cancerous cell in some location within the body. First, it must know that what it has found is not normal – it has to have some kind of memory. Second, it must somehow let other cells know that this problem has been identified. Third, the 'authorities' must send troops out to the site. Fourth, the troops must destroy the 'not normal' organism and, finally, dispose of it. Every second of our lives, these incredible processes occur inside our bodies without our awareness. Although they are far more complex, they could be described as the chemical version of sending an SMS or email, which also happens within a split second.

There are many different chemical messengers, making it possible for all the various kinds of cells to know what is happening, where to go, what to do and when to stop. There are receptors on the surface of many cells that function like satellite dishes, each one designed to 'pick up' different chemical messages.

Defective communication is the cause of many diseases, even when all the cells necessary to remove the problem are ready and waiting for instructions. A lot of current medical research concerns why the chemical messages are sometimes not sent from the site of cancerous growths (for example), and why the cells meant to receive the messages and act on them sometimes don't respond.

The role of hormones in immune functioning

Various hormones are able to profoundly affect the immune system. These include **cortisol**, which we've heard a lot about in this book as the classic stress hormone. But another important function it has is **suppressing the immune system**.

Cortisol effectively slows down cell activity. There are cortisol receptors on the surface of most white blood cells. In tissue transplant procedures (such as bone marrow, heart and lung transplants) the patient is often infused with great quantities of corticosteroids (cortisol-like hormones) to ensure that their immune system does not reject the transplanted organ. Many medications for rashes, insect bites and allergies also contain cortisone (another close relative of cortisol) as their main ingredient.

Amazingly, the side effects of cortisol closely mimic HIV/AIDS in almost all respects, including anxiety, depression and susceptibility to bacterial and viral infection. However, unlike HIV/AIDS, the immune system returns to normal when the cortisol treatment is removed.

Even more incredible, adding cortisol to T cells increases the ability of viruses to infect these cells by up to 70 per cent.

But as we'll discover in the next chapter, suppressing the immune system can sometimes be *good* for us.

> *'The stress hormone cortisol has been shown to slow down cell activity and suppress the immune system.'*

CONNECTING THE DOTS: M + B + C = MBC

Let's now pull the various components together, and show how the mind can affect the immune system. There are many biological pathways that connect the two, but the easiest to relate to are the SAM and HPAC systems.

The fight-or-flight system – FIGHT

Remember Brog, the caveman we met back in **Part 1**? When Brog encountered a threat, such as a dangerous tiger, the **SAM system** (sympathetic adrenal medullary system) kicked in to give Brog the energy either to fight, or to run away from the danger. The *sympathetic nervous system* was activated and the *parasympathetic nervous system* took a back seat.

The *other* system of importance to mindfulness and MBC covers the possibility that poor old Brog chose to fight the tiger, or didn't run away from the tiger fast enough, and got wounded in the encounter. Ah, shame … poor Brog.

For this possibility, Mother Nature evolved the **HPAC system** (hypothalamus pituitary adrenal cortex system). We could also refer to this system as the 'hopeless–helpless' system, or the 'I am wounded' system.

Oddly enough, the HPAC system is specifically designed to **suppress the immune system**. This makes no sense until you understand the paradox: **a slightly suppressed immune system results in a more rapid healing of wounds**.

When you hurt yourself, a wide range of immune-system cells rush to the scene of the wound. If this rushing to the scene is not controlled or slowed down, the swelling caused by all those eager little cells can cause even more damage to the tissue than the original wound.

For example, some people are highly allergic to bee stings or pollen. If they are not careful, they can die from being stung or exposed to pollen – not because the pollen or bee sting is 'poisonous', but because their throat swells due to the rushing in of eager-to-help cells. Their windpipe eventually closes and they die of suffocation. This swelling is nothing more than an over-eager immune system.

In *non-allergic* people, the body sets off a chemical domino effect to control such swelling, called a **hormonal cascade**. When the body realises that it's been hurt, the hypothalamus secretes a chemical called **CRF** (corticotropin releasing factor). In turn, this activates the pituitary gland to secrete **ACTH** (adrenocorticotropic hormone), which races through the blood system to the adrenal cortex, which then releases **cortisol** into the bloodstream.

As we discovered in the previous chapter, the effects of cortisol on the immune system are very powerful. This is particularly the case when it's continually secreted over an extended period of time.

In **Part 1** we saw that the SAM system no longer provides the same advantages as it did in Brog's time, and often results in chronic stress. The HPAC system is similar. It evolved to deal with wounds and **external situations** when control of a situation was lost. Nowadays, though, **internal perception** of woundedness or loss of control – i.e. emotional distress –

produces the same effects. In other words, **chronic emotional distress** – upset, fear, worry, anxiety, depression – directly results in chronic increases in cortisol. This in turn leads to long-term suppression of the immune system.

Higher-than-normal levels of cortisol have an astonishing range of effects, many of which can be directly linked to cancer, HIV/AIDS and other suppressed immune system disorders. For instance, cortisol has been shown to increase the ability of HIV to infect T cells, largely by affecting the receptors on the surface of the cells. Another study found that corticosteroids can reduce immune functioning in AIDS patients.

Once again, I'm *not* saying that cortisol *causes* these diseases. What I *am* saying is that the release of abnormal levels of cortisol through psychological stresses can certainly *contribute to the development* of these diseases.

Just to repeat: **the mere perception of control – or not having control – is enough to set off a range of profound chemical changes** in your body, all of which can help or hinder your health. As far as the body is concerned, it doesn't matter what is real and what is imagined.

Does there have to be a dramatic change in the immune system to cause disease? Research suggests there *doesn't*. It seems the immune system operates in surveillance mode most of the time – checking to see if there is anything wrong (such as the presence of cancer cells), then destroying these abnormalities on a daily basis. Even a small decrease in this surveillance activity can shift the balance – for instance, causing more and more cancer cells to slip through the net, eventually resulting in clinical cancer.

However, when the SAM and HPAC systems misfire, the negative effects are not dramatic or obvious in the short term. The research clearly shows that **acute (short-term) stress**

does not have any lasting effect on the immune system.
Your system experiences a slight drop, then returns to normal levels within fifteen minutes or so. Good news for those of us who are panicking about having a bad day!

Most of the negative effects occur as a result of a **slight decrease** in immune functioning **over a long period of time**, usually six months to two years. As a result, the focus of Applied MBC is on chronic stress and not short-term stresses.

Perception of control

Animal studies indicate that their **perceived control** (versus **actual control**) over pain is an important factor in determining which system – the SAM or HPAC – will be activated at any given moment in time. When the animal believes it has control (even when the control is not real), the SAM system is activated. When the animal believes it has NO control (even if this belief is inaccurate), the HPAC system kicks in.

For example, if a dog is trained to believe that by pushing a button in its cage it can prevent a mild electric shock, then the **SAM system** is activated, with the result that the dog doesn't develop typical HPAC-related diseases such as cancer. Even if the button is not actually connected to the electrical device, and pushing it has no effect on the frequency or strength of the electrical shocks, typically the dog does not develop cancerous tumours.

However, if the animal believes that it can do nothing at all to prevent the electrical shocks from occurring, it typically develops **HPAC-system diseases**. (Of course, these experiments need to be conducted over several months and years, rather than being short-term situations, since *short-term stress does not have a lasting impact*.)

It is the **perception of control or absence of control** that determines which system – the SAM or the HPAC system – is activated in any circumstance. This is an astonishing situation. It demonstrates what a huge difference your mind and emotions can make to your body.

The big picture is by no means complete. But scientific research has clearly shown that a wide range of psychological states lead to a similarly wide range of immune dysfunctions. And when these psychological states are long-term and chronic, the immune-system dysfunctions are equally chronic.

So, now that we've seen some of the hard scientific evidence behind the mind–body connection, we're in an ideal position to start exploring how to apply that knowledge to our day-to-day lives. That's what we'll focus on in the final part of the book, **Part 6**.

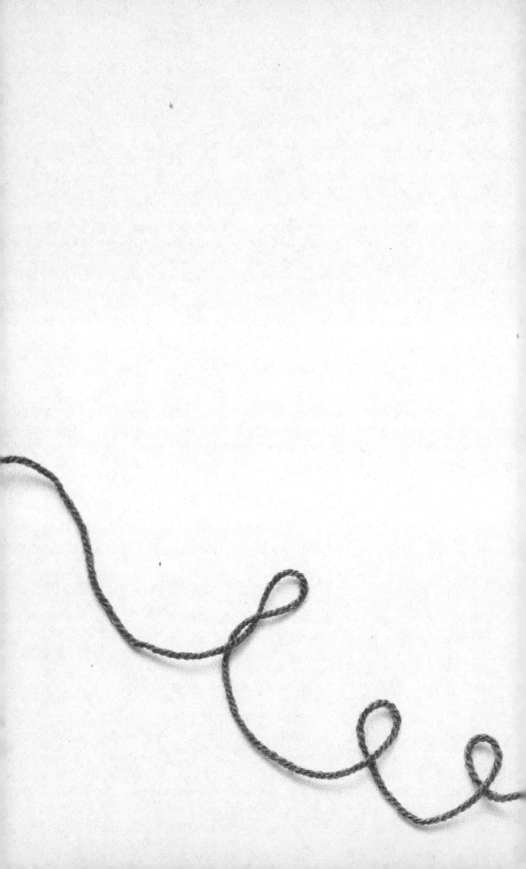

PART 6

USING MBC TO MANAGE YOUR HEALTH

WHERE TO START?

All the techniques you have learned so far in this book will help improve your overall health, but with MBC you can take mindfulness even further. By adopting the strategies set out in the following pages, you can bolster your immune system and in turn protect yourself against a range of common conditions and diseases. But where to begin?

Body work

MBC always starts and ends at the same place: the body. Yes, there will be many emotional and mental changes that happen too if you begin to practice mindfulness, but MBC is really designed to do one thing, and one thing only: **have a positive effect upon your physical health**.

Therefore, the very first thing that needs to be done is **obtain as much information as possible about your present physical condition**. So, if you haven't done it already, start by visiting your doctor. He or she may recommend getting scans, blood tests to check for important immune markers, or other medical tests to determine exactly what the condition of your body is.

It's your body, and your life. It makes sense to know what's going on, so you can make informed decisions about treatment options, both medical and non-medical. A great deal of anxiety can be lifted from your shoulders when you find out how healthy you are and what challenges you might face.

Lifestyle factors

Next you need to look at your lifestyle, from multiple angles – including personal and physical factors, relationships, work and environment – in order to work out how to allay or prevent those illnesses.

For instance, you may live in a particularly difficult *environment* that you need to take into account. If your doctor has told you you're at risk of a heart attack and you really need to relax more, but you work on a petrol rig in the middle of the ocean, this will have a big impact on your ability to take your doctor's advice. Your job is risky and high-pressure, you're isolated from your family, and these things will affect your success in applying MBC to your life.

Awareness that these factors are affecting you is a good starting point for developing mindfulness. Sometimes a situation is so overwhelming that change is absolutely necessary!

At a *personal* level, what do you like to do in your spare time? What risk factors do these activities involve that might cause you additional stress?

Most obviously, there are the *physical* aspects of your lifestyle. Are you overweight? Are you getting adequate rest and exercise? Some useful information on each of those areas can be found in the chapter on managing your physical energy in **Part 3**.

It's also vitally important, especially if you already have any kind of chronic condition, to focus on what kind of *nutrition* your body needs to stay strong.

Nutrition

Do some homework on foods and medicinal plants that may help your body, such as antioxidants and other immune-enhancing substances. Examples might include selenium – found to be very beneficial in suppressed immune conditions – vitamin C, zinc and various medicinal plants.

Be careful about using 'immune-boosting' plants and remedies. The reason is simple: rarely do such products tell you *which* part of the immune system they 'boost'. There is no such thing as a generic immune-boosting remedy. Instead, such remedies boost either **cellular immunity** (which deals with viruses and cancers), or **humoral immunity** (which typically deals with bacteria). Strengthening one of these two systems tends to weaken the other.

For example, when you're dealing with a bacterial infection, the 'immune-boosting' plant echinacea is very effective. But if you're dealing with a suppressed immune system condition, echinacea can make your illness worse.

The underlying science here is the concept of **homeostasis**, which we looked at in **Part 2** (see page 67) and the brain and body's need to balance the 'flows' of chemical responses (hormones) and electrical responses (neurotransmitters). When we are ill, these are not in balance; long-term imbalances can manifest as chronic illness.

MBC infrastructure

But as well as working to change your lifestyle, you'll also need to change the 'road' that you travel on. And it's the **mental** and **emotional** side of things – enhanced by mindfulness – that make up this road. They're the **infrastructure** behind your MBC journey.

At any point in time, life may cough up a personal or a family or an environmental or a work issue, but if you actually work on the infrastructure, you'll be able to tell yourself, 'It's a bit bumpy for the moment, but if I just get my mind into gear, I'll be able to continue on my journey until the "bumps" resolve themselves.' So in the rest of **Part 6**, we'll use lots of practical examples and exercises to focus on enhancing the 'mind' part of MBC.

BELIEVE IN YOURSELF

MBC is a collection of scientific research data and methods. We view these as the 'bricks' of Applied MBC. But what is absent from all these amazing methods and principles is the 'cement' that holds them all together: self-belief.

If you don't feel that you can stand up to the world by yourself, then perhaps you need to build up that part of you first. Do the following exercise several times (if possible), until you feel that your sense of 'Yes, I can' is stronger. This is a visualisation technique that has been proven to help the brain with anticipation and the carrying out of future actions.

EXERCISE
Building internal confidence

Step 1: Find a past experience when you felt confident.
Some time in your past you felt confident in yourself, even if it was only briefly.

Write down five such experiences. (A minimum of one is required.)

1. _____

2. _____

3. _____

4. _____

5. _____

Step 2: Select the strongest experience.
Select the strongest experience – in terms of feeling confident – from the list in Step 1.

Step 3: Recreate the feeling while practising mindfulness.
I've used colouring in here, but with some small adjustments you could use any mindfulness activity you like. (See **Parts 1** and **3** for suggestions.)

Focus your eyes on the colouring page. With your head bent slightly forward and your eyes pointed downward, remember your experience of feeling confident.

While continuing to colour in, slowly start to focus on what you were *physically doing* when this past experience occurred. Were you walking, standing or sitting? How was your body positioned?

continued

Still colouring in, focus upon your *surroundings* during this experience. *Who* was there with you? Were they behind you? In front of you? To the left or right? Was anyone saying anything?

Now move your attention to the moment when you felt the strongest feeling of confidence ('Yes, I can!'). Recreate that feeling in your body while colouring. Where in your body are you feeling it? What kind of feeling is it? Warm, hot, electric, or calm?

Step 4: Make it stronger.
Imagine that this feeling of 'Yes, I can!' has a shape, size and colour. Make the colour brighter and stronger on your colouring page and in your mind. Then make the shape of the feeling twice as large. Increase the size of the shape until your entire body is inside this shape you are colouring.

Step 5: Tap your chest.
As you feel this wonderful feeling, stop colouring, take your colouring hand and firmly – but not too hard – tap the centre of your chest, just below the top of your breastbone.

Step 6: Repeat Steps 3, 4, and 5 using the other experiences you listed in Step 1.
For each of the other experiences you listed, repeat the process of recreating the experience, feeling the feeling of 'Yes, I can!', magnifying the shape and colour of the feeling, and tapping it into your chest.

How does this help?

Probably the most important benefit of this process is that you've learned that there is a part of you that has confidence, no matter how weak your confidence feels right now. Strengthening your self-belief is a matter of finding that part of you, and bringing it into your present situation.

Mindfulness provides the vital link in this mind–brain connection.

Secondly, this confident part of you has now been strengthened, and can be accessed by gently tapping your chest again, which will bring it into the present moment.

Try it.

BE A REBEL

Another important thing you need in applying MBC is a **rebellious streak**. It's really useful to have a sceptical voice inside you that says, 'Maybe so, but ...'

We walk a very fine line in Applied MBC, as we rely heavily upon the authority of MBC science and the controlled research it produces. At the same time, we recognise the fundamental need – when *applying* this science – to promote a healthy scepticism of any external authority.

It's ironic that the most impressive 'survivors' of serious diseases – those we use as examples of how MBC can work – are also the most likely to be sceptical of any medical or scientific authority regarding their disease. They tend to be the 'difficult patients' – the ones who ask annoying questions and insist on knowing *why* the doctor wants to do certain things. Scepticism is a type of fight-or-flight response; once we resolve or accept the situation, the body returns to the state of homeostasis. This process in turn builds resilience.

When you read this book and do the exercises, don't be afraid to recognise when something doesn't work for you. Change it to suit your needs. Be a rebel when you feel the need to do so. The ultimate 'authority' in the healing process is you – not MBC, not your family, not your friends, not medical science, not some herbal remedy. It is *your* right to challenge these authorities, and establish your own.

'Waste no more time talking about great souls and how they should be. Become one yourself!'

– Marcus Aurelius (Roman emperor, 121–180 CE)

IMPROVE YOUR EMOTIONAL INTELLIGENCE

The next thing you need to know is that maintaining good health using Applied MBC is not a matter of intelligence or education. When you look at the practical exercises I outline in this section of the book, they are not technically difficult. Challenging, certainly, and probably uncomfortable a lot of the time, but not all that complicated. Most of them are logical and firmly based on common sense. You don't have to be a rocket scientist to do them.

What you need more than mental IQ is **emotional IQ**. A book like this one can teach you MBC methods and skills. But personal qualities can't be taught – they need to come from inside you.

What is emotional IQ? It's the ability to allow yourself to feel what you are feeling, *know* what you are feeling, *value* and *trust* what you are feeling, and *express* what you are feeling.

When you have a high emotional quotient (EQ) you are not just adept at acknowledging and expressing your feelings. You are also capable of having effective insights into the behaviour of others and choosing your actions wisely to influence outcomes.

Most of us spend more time than we would like thinking about when we lost control or misunderstood other people. To be mature and effective in the company of others, we need to connect with our inner self and our uniqueness.

From a neuroscience point of view, emotional intelligence helps the body's homeostasis systems (see page 67) to function more efficiently and effectively, in both hormonal and neurotransmitter operations.

Key learnings for emotional intelligence

We all know that changing behaviour in a sustained and genuine way is extremely difficult. But here is a list of behaviours to work on in order to gain the maximum benefit from Applied MBC. (Chances are you're working on a few of these already.)

1. Actions speak louder than words

To earn other people's trust and respect, take your own word seriously. Do what you say you will do. Make choices, admit to them and stick to them.

2. Listen well

If you want to affect people positively, then try by listening with 100 per cent of your attention.

3. Show that you care

Though we think our friends and loved ones know how much we care, they usually don't. We all want to be told and shown – *often*.

4. Sometimes it's what you *don't* say

You don't always need to talk. Sometimes others will appreciate you more for what you *don't* say.

5. Give people space

Be sensitive to the needs of others. Sometimes the best way to get along with someone is *not to be there*.

6. Express appreciation to others

This is a powerful ritual that fuels positive emotions and is as beneficial to the giver as to the receiver. The more detailed and specific, the higher the impact.

7. Get the game rules right

To get results from others, tell them, 'This is what I want. If it's not done these are the consequences.' Check that they understand and agree. Follow through and make sure they do what you've asked.

8. Show respect

When you're dealing with angry people, facts don't work. Care and respect *do*. Listen, empathise and show respect.

9. Ask questions first

When negotiating with someone, make it your policy to ask questions first. When you question people, you invite them to think along your lines, which is more tactful and successful than *telling* them how you think.

10. Avoid arguments

Scrap the notion that says 'If someone disagrees with me, my job is to change their mind.' Instead, try the philosophy 'If someone disagrees with me, my job is to let them do so.' Telling people 'You're wrong' is a great way to make enemies. Admitting you're wrong or agreeing to disagree can be a great way to start friendships.

11. Don't get angry or take offence

Only little people make nasty remarks, and only little people take offence. *Be a big person.*

12. Learn to control your anger

When you're angry, you have an expectation that's not being met. Ask yourself if the expectation is realistic. Try to see things from the other person's point of view. Put it in perspective: how important will this situation be a year from now?

13. Anger doesn't motivate

Because we sometimes delay action until we reach screaming point, we may believe that it's the screaming that motivates people. It isn't. People take us seriously because of what we *do*, not because of how loudly we scream.

14. Give people the benefit of the doubt

Sometimes people let you down. You have a choice in how to respond: you can criticise and humiliate them or you can attempt to fix the problem. *Praise* before criticising. *Remind* rather than tell people. Admit that *you're* not perfect either. Look to the future rather than blaming others for the past. The golden rule is: give people the benefit of the doubt.

15. Mistakes – the hotpot of learning

A relationship is like a business. It's either getting better or getting worse – there's no standing still. *If things aren't improving, then you are living without learning.*

PERSONALITY HARDINESS

To help you deal with stressful events in life, including illness, it's important to develop resilience, so that you can persist and have the stamina to get through major challenges and setbacks. The key to this, I feel, is a concept called **personality hardiness**. This idea was developed by Suzanne Kobasa, a prominent researcher and theorist in the area of how psychological factors affect health.

Basically, personality hardiness is a collection of characteristics that help you deal more effectively with stressful life events. There are three main components:

1. **Challenge**

2. **Commitment**

3. **Control**

Say, for example, you've been ignored, criticised, humiliated, rejected and hurt several times in your life, by several important people. There are two ways of looking at these 'facts':

- You can use them as evidence that you are not worthy of anyone's attention – you are not good enough, and your life is not worth living.

- You can view these same 'facts' as proof that you are strong enough to survive without anyone else's approval!

Which option do you prefer?

The problem with stressful events is not **what happens**. The problem lies in **how you respond**.

When you throw a 'brick' (a painful experience) at some people, they get knocked down and become afraid of standing up again, in case another brick gets thrown at them. They come to see bricks as sources of pain.

Other people will catch the brick and *build* something with it. They'll see bricks as resources to create something valuable. They'll say thank you for these bricks, even if they occasionally get knocked down by one!

What have you done with the bricks that have been thrown at YOU? To quote a famous saying (source unknown): 'When life gives you lemons, do you throw them away or do you make lemonade with them?'

Perhaps your response is: 'That's all well and good, but if you understood how many times I've been hurt, you'd see why this is unrealistic for me.'

The irony is that the more 'bad' things that you have in your past, the greater your potential to build hardiness! After all, you've already managed to survive quite a few bricks, haven't you?

This is a whole new way of looking at life: **the more pain you have in your past, the more building material you have to create a sturdy future**.

Somehow you were able to survive and moved past these difficult experiences. So when another difficulty arises, you can take the attitude: 'I don't necessarily know how I'm going to get through this, but I know from past experience that I'll get through it somehow.'

Hardiness uses negative experiences to strengthen. This isn't an *intellectual* concept; it's a *gut-level* reaction to circumstances. This notion of hardiness coincides with my own observations of long-term survivors of life-threatening disease.

Once again, **it's not your past that is the problem – it's what you've done with it that may be the problem**. The challenge is to get yourself from 'I believe it's **possible**' to '**I know** I can!'

So, getting back to those three components I mentioned, let's look at each one in turn.

Challenge = 'I want to'

If your life was easy, with no sense of adventure, and no goals to work hard for, your body would not be very happy. This is because the sense of **challenge** – taking risks, getting excited about trying to achieve something – causes various hormones to be released, such as human growth hormone. (The *absence* of challenge can have the reverse effect. For example, the wasting away of muscle tissue in some HIV/AIDS cases, known as the HIV wasting syndrome, is partially caused by abnormally low levels of human growth hormone. We'll look more at this issue of challenge in the next chapter.)

The first quality you need to find inside yourself is this sense of interest in getting somewhere, the determination and excitement to achieve something.

EXERCISE
Challenge = 'I want to'

Write a brief summary of five events or periods in your life when you felt this sense of excitement. You wanted to get out of bed early because you had exciting things to do.

These events or periods in your past do not have to be dramatic. They can be simple things such as the excitement of learning to ride your brand-new bicycle, or the excitement (and nervousness) of arriving at work for the very first time. It's the feeling of excitement and challenge that is important, not the details of the event itself.

The key idea is that these events represent times when (a) you were excited (and perhaps nervous too); and (b) you felt a sense of challenge – you were not quite sure how you were going to achieve what you'd set out to do, but you were willing to risk trying.

_____

_____

_____

_____

Commitment = 'I will'

While challenge is based on the thrill of 'I don't quite know how I'm going to do this, but I'm going to try', **commitment** is more focused upon the determination to persevere, and the sense that something is really worthwhile.

EXERCISE
Commitment = 'I will'

Briefly summarise five events or periods in your life when you felt this sense of commitment. You were determined to do something, even though you didn't necessarily know how you were going to achieve it.

Again, these events or periods from your past do not have to be very dramatic. They can be simple situations in which you made a decision to do something, then did it. For example, the actions you took to get your driver's licence, even though you were terrified of the tests. This would demonstrate your determination (commitment) to achieve what you wanted.

_____

_____

_____

_____

_____

Control = 'I can'

This simply means that you're able to start or stop something, or change the direction a situation is moving in. You may not know *how* you can do this, but you're confident that you can if you want to.

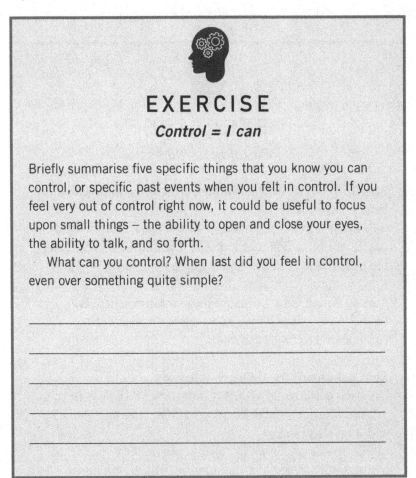

EXERCISE
Control = I can

Briefly summarise five specific things that you know you can control, or specific past events when you felt in control. If you feel very out of control right now, it could be useful to focus upon small things – the ability to open and close your eyes, the ability to talk, and so forth.

What can you control? When last did you feel in control, even over something quite simple?

_____

_____

_____

_____

_____

Building hardiness

In the earlier exercise, 'Building internal confidence' (page 173), you experienced how you can take a small feeling and make it much stronger. You're now going to use exactly the same method to strengthen the three parts of hardiness that you've just been working on – challenge, commitment and control.

EXERCISE
Building hardiness

Step 1: Select an experience from each of the three previous lists.
Select *one* experience from *each* of the three lists you made in the 'Challenge', 'Commitment' and 'Control' exercises. (Select the ones that will be easiest for you to remember.)

Step 2: Select the strongest of those three experiences.
Of the three past experiences, which is the easiest to remember? Start with that one.

Step 3: Recreate the feeling in your body.
I've used colouring in here, but with some small adjustments you could use any mindfulness activity you like (see **Parts 1** and **3** for suggestions).

Focus your eyes on your colouring page. With your head bent slightly forward and your eyes pointed downward, remember the experience of feeling challenged, committed or in control (whichever is applicable).

While continuing to colour in, slowly start to focus on what you were *physically doing* when this past experience occurred. Were you walking, standing or sitting? How was your body positioned?

continued

Still colouring in, focus upon your *surroundings* during this experience. *Who* was there with you? Were they behind you or in front of you? To the left or right? Was anyone saying anything?

Now move your attention to the moment when you felt the strongest feeling of being challenged, committed or in control. Recreate that feeling in your body while colouring. Where in your body are you feeling it? What kind of feeling is it? Warm, hot, electric, calm?

Step 4: Make it stronger.
Imagine that this feeling has a shape, size and colour. Make the colour brighter and stronger on your colouring page and in your mind. Then make the shape of the feeling twice as large. Increase the size of the shape until your entire body is inside this shape you are colouring.

Step 5: Tap your chest.
As you feel this wonderful feeling, stop colouring, take your colouring hand and firmly – but not too hard – tap the centre of your chest, just below the top of your breastbone.

Step 6: Repeat Steps 3, 4 and 5 with the other two strongest experiences that you listed in Step 1.
Repeat the process for the strongest experience in each of the other two categories. For each of the other two experiences you listed, recreate the experience, feel the challenge, commitment or control, magnify the shape and colour of the feeling, and tap it into your chest.

Step 7: Select another three experiences from your lists.
Select another experience from each of the three lists.

Step 8: Repeat Steps 2 to 7 for the rest of your experiences.
In groups of three – one from each list – repeat these steps, until you have strengthened all fifteen experiences.

Keep working on these exercises until you develop the hardiness to use the *bricks* life has thrown at you to build the road you need to travel on in your MBC journey.

> 'Things turn out best for the people who make the best out of the way things turn out.'
>
> – Art Linkletter (Canadian radio and TV personality)

CREATE A COMPELLING FUTURE

One of the major objections to MBC interventions is that only about 50 per cent of participants benefit. These programs must be unreliable, the argument goes. Why would MBC work for one person, and not another?

There may be many reasons why MBC isn't working for you. Perhaps you're not following the advice of your **doctor** or **alternative practitioner.** Maybe you're not getting enough **exercise** and **rest**, or **eating** right. Do you **believe in yourself**? Are you confident and free of guilt about your illness? Are you a bit of a **rebel**, constantly questioning what you are told? Do you have a high **emotional IQ**? And do you have the **hardiness** to persevere with MBC, even when things get tough?

You may be working on all those areas, and reaping the benefits. But I've observed one final reason why some people don't succeed with MBC: they have feelings of low self-worth because they don't have a **compelling future** to move towards. This is a fundamental obstacle to MBC effectiveness.

From an MBC perspective, there are **two main reasons** for imagining some kind of desirable future that you can work towards:

1. Challenge, pleasure and interest have a specific range of hormonal consequences, including increases in adrenaline and human growth hormone. They also result in the release of various neuropeptides, all of which have the ability to improve a suppressed or over-active immune system, among other effects.

2. A person with no future-related goals has very little reason to be working to improve their health.

Away from pain: An unreliable motivation

When a group of people in roughly the same physical condition start a health program – diet, exercise, yoga, meditation, whatever – the group soon splits into two subsets. The first subset thrive and get healthier, and this continues over time. The second subset get healthier for a while, then the effects fade and they get sick again.

Why would the same program have such different effects? The simple answer is that some people want to get healthier **because they have a future they want to get to**, while the other people want to get healthier **because they are afraid of pain and illness**. In other words, some people are motivated **away from** something they fear (pain, illness, death), and others are motivated by desire to move **towards** something exciting. These two groups are doing exactly the same things, but for completely different reasons.

Running *away from* pain and illness is not the same thing as running towards health and aliveness. When you say, 'I'm sick and I want to get better', you're really saying nothing more than 'I want to move from where I am.' But if I were to ask you *why* you want to move, you'd say one of the following. 'I want to move because I prefer to be somewhere more pleasurable' (in other words, you want to move towards *something better*), or 'I want to move because I don't like it here' (in other words, you have *no specific destination*). When you're motivated by avoiding pain and illness, your mind is firmly fixed on what you *fear*. (Part of this is the fear of *not having enough time*, which we looked at in **Part 1**.)

If these fears are your central focus, health and happiness won't feature in your mental and emotional state. And when a pain or illness first subsides, your motivation will decrease, but only until it resurfaces. Therefore, this 'away from' motivation is not constant, but comes and goes.

In a fascinating study, Professor Stanislav Kasl and his colleagues examined a specific type of military cadet. These cadets had over-achieving fathers (defined as those who had achieved more than their education suggested they would) and high levels of motivation, yet they were not performing as well as they thought they should.

This is a classic **stressed power syndrome** situation. (We encountered this syndrome briefly in **Part 1**.) It's interesting to note that first-born children often have this syndrome, because first-time parents tend to invest all their hopes and dreams in this first child, and can create the impression that they only approve of (love) the child when he or she achieves something worthwhile.

Kasl and his colleagues found that this group of cadets were more likely to develop clinical symptoms of infectious mononucleosis (glandular fever or 'kissing disease') than other cadets who did not have this 'I must succeed, but I might fail' motivation. Though cadets in both groups became infected with the same virus, those with the stressed power syndrome were more likely to get ill as a result. This suggested that their immune systems were less effective in dealing with the infection.

Of course, anyone can get glandular fever. But when you're infected with it your psychological make-up can affect how ill you become.

In people with stressed power syndrome, **fear of failure** is the primary motivating force, rather than a desire to succeed. Consequently, a person with stressed power syndrome is constantly on the lookout for any threats to their success. They experience failure as devastating, and a clear indication that they are of little or no worth. This becomes a major problem when their expectations are higher than their abilities, as witnessed with the military cadets.

It's not surprising that type A personalities with a 'must succeed' (rather than 'want to succeed') motivation drive are attracted to competitive environments. But in these environments their drive to succeed can cause a constant over-activation of the SAM system, leading to heart conditions, high blood pressure and other SAM-related conditions.

Towards a goal: A more reliable motivation

When you are moving *towards* health and happiness, you keep your mind firmly on where you are going. The motivational force is constant – regardless of pain or absence of pain, health or disease – until you reach your goal. This motivation makes a major difference to your immune system. This is a big reason why the same medical treatment or healing method can work for one person, and not for another.

Most people who are ill are clear on the fact that they do not want to be ill. However, how many people know what they want to move *towards*? What kind of future do you *want*? This is *not* the same as asking what you *do not want*!

If there are no goalposts on a soccer pitch, what will the players do? There they are, running with the ball, knowing that they have to move from one end of the pitch to the other, but with no idea of what to do when they get there! *They can't score because there aren't any goalposts.* After a while, they'll realise that the game is silly, so their choices are to kick the ball around aimlessly or walk off the pitch.

Life is much the same – you need goalposts, otherwise the game has no real meaning or purpose. And just like in a game of soccer, after you score the first goal, there's nothing to stop you from scoring another, then another.

What is a compelling future?

Dr Bernie Siegel, one of the first medical practitioners to publicly acknowledge MBC, asked people who were dying of cancer why they wanted to stay alive. Most said they needed to go on living due to their obligations to their husband or wife, kids, other family and friends.

So, he tricked them. He asked family and friends to gather around the dying person, then he proceeded to delegate the person's obligations to the people around the bed.

Once all the obligations had been dispersed, a very strange thing happened. The patients began requesting Dr Siegel's cancer-intervention model. Simply put, once their *obligation* to live had been removed, their *passion* to live returned. A classic case of 'want to' versus 'have to' motivation.

This is an extreme example, but many of us put aside our passions and live a life based only on obligations. Yet to make the most of the MBC, you need to create a compelling future. For the truth is that, until such time as you have filled your own cup, what do you have to give to someone else?

So, what exactly is a **compelling future**?

First of all, it usually contains a range of goals, both short-term and long-term. These goals may be silly and irrational, or the opposite; it doesn't matter. What matters is that you experience excitement and interest in reaching for them, and pleasure and fulfilment in achieving them.

'But,' I hear you say, 'isn't that shallow and superficial? My sense of obligation to loved ones isn't "exciting", but at least it's meaningful – I can make a difference to other people's lives!'

If that kind of thought has gone through your mind, you are partially correct. For most people who are ill (and those who are not), life is a mixture of work and family obligations that may or may not bring a sense of fulfilment. For many people the main fulfilment in life is to be found in relationships – raising children, an intimate relationship, close friendships. It's not surprising that, for most of us, happiness is locked into a rather small group of people. Other areas of potential pleasure and fulfilment for many people are work and study. What else is there?

Surely there has to be *more* to life? But most people have given up on trying to find that 'more', and have settled for a reasonable, OK existence. When a life-threatening disease comes along, what compelling passion is there to counteract the temptation just to give up and escape from a painful, monotonous life?

Studies of children in orphanages show that lack of stimulation and interest results in deterioration of mental faculties and health in general. The same outcomes are found in retirement homes. The human species requires challenge and stimulation to thrive. Boredom is as dangerous as a life-threatening disease. **Fundamentally, you need challenge and stimulation as much as you need food and water to survive**.

Research by Dr Marian Diamond at the University of California clearly supports this. She found that, when isolated and just provided with the basics for survival (food, water, warmth) but no interaction with others, rats have a much shorter lifespan than those that are constantly playing with each other and are provided with new toys every few days. Clearly, having interesting things to do gives the rats a reason to live longer – up to 50 *per cent* longer than the average rat's lifespan.

The six practical steps below will help you start to create this **compelling future** for *yourself*.

1. Increase the simple pleasures in your life
Savour the small stuff.

EXERCISE
Simple pleasures

Answer the following questions as honestly as possible:

- What is your favourite taste? When last did you taste it?
- What is your favourite smell? When last did you smell it?
- What is your favourite music? When last did you hear it?
- How many times have you laughed today?
- Have you allowed someone to touch you gently today?
- How much pleasure have you had today? Be specific about what it was that you enjoyed.

We often get caught up in the big, important 'meaning of life' stuff, and forget about the small things that fill our world with pleasure and joy. Perhaps you're not ready yet to examine the 'big issues' of purpose, passion and a compelling future. That's OK – on one condition: make sure you have a good day.

Laugh. Watch a movie that you love. Eat your favourite food. Ask someone to give you a gentle massage.

Start to enjoy being in the moment. Only then will it make any sense to work towards enjoying more of this thing we call life.

2. Find beauty where you are

Sometimes we get so caught up in the pain, suffering and seriousness of life that we completely forget about what we actually have, right here and now.

EXERCISE
Finding beauty

Sit still – anywhere – and just look around you. Find something – anything – that you consider beautiful. It doesn't have to be anything profound or amazing – just something that you haven't previously paid much attention to, such as a flower, the clouds, a picture, your own hands.

Don't think about what you are looking at – just look at it, and appreciate it. This could be the only time you will ever see it. If that makes you sad, that is also OK.

3. Let something go, and let something new in

This is a simple thing to do.

EXERCISE
Letting go, and letting in

Look around your home. How many books, papers, objects, pictures and pieces of furniture have been sitting in exactly the same place for a long time? How many drawers and cupboards have had the same things piled into them for ages?

Give your home a clean. Throw or give away all the things you haven't needed or used for a long time. Give old books and magazines away to someone who could use them. Throw away or burn old bills that have been sitting there for years. Move a few pictures on the walls. Turn the mattress over.

Now bring something new into your home, even if it's just a second-hand book that you want to read.

---

4. Get your affairs in order

Relax – this is not about anticipating death, it's about **freeing yourself up to live**.

EXERCISE
Getting your affairs in order

Sort out your finances, pay your taxes. Ensure that your will is current. If you haven't got one, make one. If possible, give things away to your loved ones now, so that you can enjoy their delight in having them. Write down what you want to happen at your funeral.

Yes, all this sounds bizarre, but until you get these issues out of the way, they will always be on your mind. Right now, you need to move beyond them so you can focus on living.

Then do something *fun* – something you've wanted to do for a long time, but have always put off. After all, once you've put your death behind you, what else is there to do but *live*?

David went to a funeral home and tried out all the coffins. The funeral director was horrified. But when David found a coffin he liked, he paid for it, then gave the funeral director instructions for his funeral, and paid for that too. The same day, he went to an estate agent to buy a new house. He signed the papers the next day.

You don't have to go to those extremes, but it's really important that you get your thoughts and plans regarding your death sorted out, so that you can get on with life.

5. Work out what you do and don't want

What is the compelling future you want to move towards – an exciting and powerful reason to get out of bed each morning? To help you decide, let's look more deeply into how people are motivated.

EXERCISE
Things you do and don't want

List twenty things you *don't want* in your life.

1. _____
2. _____
3. _____
4. _____
5. _____
6. _____
7. _____
8. _____
9. _____
10. _____
11. _____
12. _____
13. _____
14. _____

continued

15. _____
16. _____
17. _____
18. _____
19. _____
20. _____

Now reverse the process and find twenty things you really *do* want.

1. _____
2. _____
3. _____
4. _____
5. _____
6. _____
7. _____
8. _____
9. _____
10. _____
11. _____
12. _____
13. _____
14. _____
15. _____

continued

16. _____
17. _____
18. _____
19. _____
20. _____

Did you notice how easy it was to compile your 'Things I don't want' list? But what about your 'Things I do want' list? Did you notice that by the time you reached point 6 or 7, you were starting to struggle?

For the most part, we as humans are very sure of what we *don't* want in life, but nowhere near as certain of what we *do* want. We are clear on what we want to move *away from*, but are vague and wishy-washy about what we want to move *towards*. The key to moving towards what you do want is to create a very clear picture of what you want your future to be. If you *don't* know where you're going, then any road will get you there!

Now it's time to put each 'want' item through **a quick test**, to determine if it's truly compelling. Ask yourself: 'When I think about this "want", do I feel a physical sensation of excitement?'

The body never lies. A good idea is not the same as an exciting and pleasure-creating one. Your idea must cause some physical sensation – a tingle, smile, a blush – anything to indicate that your body is saying, 'Ooh, yes, that would be nice!'

When I ask someone what they want in life, I typically hear a long list of rather good ideas. However, if their body does not get excited – animated hand movements, eyes sparkling, flushed cheeks, smiling, laughter – I essentially ignore what they're saying.

Listen to your body – it has no reason to lie to you. But your mind can, and will. As we saw earlier in the book, the unconscious mind cannot tell the difference between what is real and what *seems* real – what you want and what you *think* you want.

You can work at MBC as hard as you want, but nothing will change until you see yourself as healthy and motivated *towards*, not *away from*, something. Therefore, any goal that contains the words 'stop' or 'not' needs to be rephrased so that it clearly reflects what you *want*, not what you *don't want*; what you want to *start*, not what you want to *stop*; and so on. In other words, it needs to be stated in the positive, not the negative.

To take an extreme example, saying 'I don't want to die' means that the picture in your head is one of suffering. This picture has to be pretty scary to motivate you to do something. Now try turning it around. Instead of saying 'I don't want to die', say 'I want to live.' The new picture is one of aliveness, not suffering.

But if you want to do it properly, you'll need to be specific and state *why* you want to live: to fulfil some exciting goal. For instance: 'I want to live so I can create the most beautiful paintings the world has ever seen, and I want to paint lots of them!' Now you're motivated by an even more powerful picture of yourself passionately involved in something you really love.

When you've turned the negative picture around into an exciting and pleasurable one, you'll notice that you want to start moving towards that exciting future RIGHT NOW! Furthermore this new outlook is very likely to help improve your health too – by increasing positivity, reducing anxiety and, in turn, strengthening the immune system.

6. Confront your doubts and fears

Generally speaking, the minute you hit on something that you really want to do, **fears** will emerge. A whole range of objections, and thoughts of 'What will people say?' and 'Don't be silly! You can't do that!', will flood your mind.

This is probably the scariest stage of the process, because you get slammed from the inside (yourself) and outside (other people) with all sorts of quite understandable objections.

Don't start thinking you should be following *someone else's* idea of what *should* make you happy. No matter how much someone else loves you, they can't experience your pain *for* you. They can't step into your body and actually *know* what it's like. Which makes *you* the one and only authority.

The best response is to recognise that what you want is *not* understandable or meaningful to anyone else, and that you're not doing it to make *anyone else* happy; you just want to enjoy something for the sake of the pleasure it brings *you*.

We've already seen that dealing with fears is very important in terms of the immune system. And fears are the most powerful sabotage mechanism imaginable.

Notice what fears and doubts emerge as you think about your goals, and write them down. If you aren't honest about your fears, you won't be able to deal with them.

How often have you thought that it would be wonderful to do something – such as the exercises in this book – but you just don't feel you have the time or energy? Have you considered the possibility that this feeling is nothing more than fear? What do you fear might happen if you actually took control of your body and your health?

Remember the compelling goal from earlier? 'I want to live so I can create the most beautiful paintings the world has ever seen, and I want to paint lots of them!' Even if you are already ill, if your dream is to paint, then start art classes! Don't wait until you get better – start acting on those dreams **right now**! Even if you're tired and in bed, start painting! **Just do it!**

It's important, though, to bear a couple of things in mind.

Don't expect to take the first small step and arrive at instant gratification. Life is a series of little steps and big steps. The *biggest* step is the first little step, even if it's just a phone call to find out what art classes are available. Have a goal, try something, stop, look, correct, act again. If no art classes are available, then phone again later, or ask somewhere else. Just keep on going.

And one more simple rule: where your thoughts flow, your energy goes. Don't just orientate your *thoughts* towards what you want, but orientate your *behaviour* in the same direction.

So, go on, start – do it! What do you have to lose? Have some fun. The world is waiting for you to deliver your gift, so give it!

To get started:

1. Increase the simple pleasures in your life.

2. Find beauty where you are.

3. Let something go, and let something new in.

4. Get your affairs in order.

5. Work out what you do and don't want.

6. Confront your doubts and fears.

PUTTING IT ALL TOGETHER

Now you have learned to use MBC to build your self-belief, resilience and positivity. Return to these MBC exercises again and again, until you feel you have mastered them and strengthened yourself psychologically. At the same time, complement this personal development with your regular mindfulness activities.

Over the months, you will build a sense of calm, purpose, focus and contentment. This in itself will be a wonderful thing. But it's not all you will gain. Given everything we have learned about mindfulness and MBC, you can be confident that your new mindful approach to life will pay dividends not only in a greater sense of physical wellbeing but also potentially life-saving improvements in your body's defence systems.

So whether you're just a little stressed, feeling run-down or chronically ill, don't put it off – start tapping into the amazing powers of mindfulness and MBC **TODAY**!

AUTHOR'S NOTE

Over 40 years ago, Dr Ainsley Meares, a psychiatrist, wrote about relaxation, meditation, the mind and the brain. He was one of the first to really explore the mind–body connection, and his work inspired me as a young psychologist and scientist. My own journey, particularly as scientific revelations have multiplied, has been all the richer because of this 'big thinker' of the past.

In his book, *The Wealth Within* (Hill of Content Publishing: 1985), his epilogue beautifully captures what we have now come to know as mindfulness and the mind–body connection.

> We have talked of life, and many doubts have cleared from my mind. But it is only the doing that counts.
>
> First let us guard and strengthen our body for it is the fortress in which we dwell, and from which we must fight.
>
> Let us free our mind. Temper it with discipline, and enrich it with knowledge, for our mind is the essence of our being.
>
> Calm comes to us. The calm and the stillness amid the clamour and the action. It is the calm of the spirit.
>
> We understand beyond the constraints of logic, and our mind is free to range from well-worn paths of the orthodox.
>
> Secure when silence comes about us, yet rejoice in the company of our fellows, so that we need seek neither the solitude of the hills nor the merriment of the games and eating houses.

We work to contribute to the land of which we are part, and to maintain our self that we may add to the well-being of those around us. And we enjoy the restorative power of leisure that we might do these things better.

When loves comes it purifies us, and in the act of love we transcend the earthly and so enhance our being.

Our mind is clear. We see the colour of it all and the meaning behind that which we see.

When we understand, there are no opposites. They have merged in the greater picture about us.

We know of pain and grief, but our mind is still and there is no hurt in it.

The seasons come and go. The planting, the ripening and the harvest. The birth and growth and death. We feel the rhythm and the harmony of it all. And it is good.

And what of this other thing that comes in the eye of the storm and in the stillness of night, yet resides in a drop of dew? Cherish it, for it is born of the spirit and transcends all else.

More mindfulness tools from Stan Rodski

Colourtation – *the New Meditation*

The simple act of colouring has the power to reduce stress, improve mood and kindle creativity by creating new neural pathways and connections in our brains. Stan Rodski's *Colourtation* series spawned a colouring-book boom and was included in Oprah Winfrey's 2016 Christmas Wish List.

'Filling in the blanks has become one of my preferred ways to de-stress. Who knew? Besides being just plain fun, these three [books] will help with mental agility, focus and inner peace.' Oprah Winfrey

Published by Hardie Grant Books. Available at all good bookstores and at www.colourtation.com

Anti-Stress colouring books

The designs in these six colouring books are based on Stan Rodski's research into neuroscience, particularly the beneficial effects of using repetition, pattern and control to quickly relax the brain. Offering fascinating insights into how we can better understand and improve our brain health, the series also looks at how colour affects our feelings, how to stimulate the brain, our fight-or-flight response, the science of focus and attention, and how to improve our thinking agility and resilience.

Available at www.colourtation.com